Synthesis Lectures on Wave Phenomena in the Physical Sciences

Series Editor

Sanichiro Yoshida, Dept of Chem & Physics,SLU 10878, Southeastern Louisiana University, Hammond, LA, USA

The aim of this series is to discuss the science of various waves. An emphasis is laid on grasping the big picture of each subject without dealing formalism, and yet understanding the practical aspects of the subject. To this end, mathematical formulations are simplified as much as possible and applications to cutting edge research are included.

Sanichiro Yoshida

Fundamentals of Optical Waves and Lasers

 Springer

Sanichiro Yoshida
Department of Chemistry and Physics
Southeastern Louisiana University
Hammond, LA, USA

ISSN 2690-2346   ISSN 2690-2354   (electronic)
Synthesis Lectures on Wave Phenomena in the Physical Sciences
ISBN 978-3-031-18190-0   ISBN 978-3-031-18188-7   (eBook)
https://doi.org/10.1007/978-3-031-18188-7

This Springer imprint is published by the registered company Springer Nature Switzerland AG
The registered company address is: Gewerbestrasse 11, 6330 Cham, Switzerland

This book is dedicated to my wife, Yuko Yoshida, and my parents, Sonoko and Akira Yoshida.

Preface

This book stems from my lecture notes and research logs. Over the last four decades, I conducted research projects in many subfields of optics. I also taught various physics courses for related subjects. During this period, I have found numerous physical meanings in what I observe in my research and in the material I teach. Some of these observations are what I had taken for granted for a long time. Finding the physical meaning of a phenomenon is joy and, at the same time, leads to new ideas. I decided to summarize these daily findings in a book.

The target audience of this book is undergraduate students majoring in physics or physical sciences and graduate students in engineering. The primary aim of this book is to help the reader grasp the basics of optical waves intuitively but based on the underlying physics. For more advanced information, the reader is encouraged to consult references written at a more detailed level on the subjects.

We learn subjects most efficiently through research and teaching because we need to analyze and digest the underlying concepts to conduct research or teach. I have tried my best to provide comments and multiple interpretations on the topics covered by this book. That is why sometimes the descriptions are long, and the derivation of the governing equations is detailed. If the reader does not need those comments or mathematical procedures, I encourage them to skip these parts.

Discussions on the observations and findings with colleagues, teachers, and students are essential for the learning process. I cannot thank enough people who helped me understand various concepts through discussions. I owe a great deal to my high school and college teachers because my understanding always originates from what they taught me decades ago. I am grateful to my parents for providing me with the opportunity to receive such an excellent education. William Coleman, my colleague and research partner, gave me critiques, which I highly appreciate. I am thankful to Paul Petralia of Springer Nature for his help during the writing process of this book. Finally, I thank my wife, Yuko Yoshida, for her continuous support.

Hammond, LA, USA   Sanichiro Yoshida
August 2022

Contents

Review of Wave Dynamics

1.1 Oscillation and Wave

A wave is the propagation of energy. Energy is the product of force and displacement. These statements indicate that a wave consists of force and displacement components. Let's discuss the wave phenomenon accordingly. We start the discussion with a mechanical system as it is more intuitive than an electromagnetic system.

1.1.1 Spring–Mass Systems

Oscillation involves a force that causes the oscillating entity to return to the center of the oscillatory motion. The general term for such a force is recovery force. Elastic force takes the most basic form of recovery force. The amplitude of an elastic force vector is proportional to the displacement from the equilibrium position (the center of oscillation), and the direction is opposite to the displacement vector. This relationship between force and displacement is known as Hooke's law [1, 2].

Harmonic oscillation of point mass

A spring–mass system [3–5] is a simple and informative physical system to discuss oscillation due to elastic force. Figure 1.1 illustrates a sample spring–mass system. Here the object connected to the right end of the spring is considered a point mass. The middle illustration represents the moment when the mass passes through the equilibrium point from the right. The equilibrium point is where the spring is at its natural (unstretched) length. The top and bottom illustrations illustrate the situation before and after the mass passes through the equilibrium point. The arrows represent the elastic force exerted by the spring.

© The Author(s), under exclusive license to Springer Nature Switzerland AG 2023   1
S. Yoshida, *Fundamentals of Optical Waves and Lasers*, Synthesis Lectures on Wave
Phenomena in the Physical Sciences, https://doi.org/10.1007/978-3-031-18188-7_1

Fig. 1.1 A mechanical system consisting of a single spring and point mass

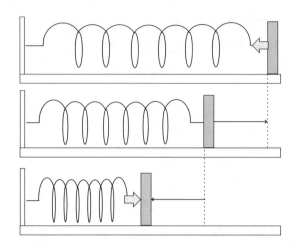

Consider the situations depicted by the top and bottom illustrations. Since the magnitude of the elastic force is proportional to the displacement, the acceleration increases as the mass moves away from the equilibrium point. This feature causes the velocity of the mass to decrease after passing the equilibrium point. Eventually, it becomes null. At that point, the acceleration is still toward the equilibrium point, causing the mass to start moving backward with zero initial velocity. In other words, the mass switches its direction after a momentarily stop. The point at which the mass makes this stop is called the turning point.

After some time, the mass returns to the equilibrium point from the turning point, passes the equilibrium point due to inertia, and moves toward the other turning point. By the same dynamics as at the first turning point, the mass switches its direction of motion after a momentary stop. In this fashion, it keeps oscillating back and forth around the equilibrium position.

Equation of motion

Considering velocity damping force (such as air resistance or friction), we can express this oscillatory motion with the following equation of motion [4–6]:

$$m\frac{\partial^2 \xi(t)}{\partial t^2} = -k_{sp}\xi(t) - b\frac{d\xi(t)}{dt} \qquad (1.1)$$

Here $\xi(t)$ represents the displacement of the point mass from the equilibrium position, m is the mass of the point mass, k_{sp} is the spring constant, and b is the velocity damping coefficient.

For simplicity, introduce the following parameters β and ω_0:

$$\omega_0 = \sqrt{\frac{k_{sp}}{m}} \tag{1.2}$$

$$\beta = \frac{b}{2m} \tag{1.3}$$

Here ω_0 is referred to as the natural (angular) frequency and β as the decay constant. It will shortly become clear why these parameters are called in these ways.

Dividing both-hand sides by m and using Eqs. (1.2) and (1.3), we can rewrite Eq. (1.1) in the following form:

$$\frac{d^2\xi(t)}{dt^2} + 2\beta\frac{d\xi(t)}{dt} + \omega_0^2\xi(t) = 0 \tag{1.4}$$

Equation (1.4) is a linear differential equation [6]. We can find the displacement of the point mass as a solution to this linear differential equation. Note that this differential equation has no source term (the right-hand side of the equation is zero.) Hence, Eq. (1.4) is classified as a linear homogeneous differential equation. Physically, this corresponds to unforced oscillation. In other words, there is no driving force in the system. Being represented by a linear differential equation, the system is called a linear system.

When a driving force is acting on the mass, the solution $\xi(t)$ represents forced oscillation. In this case, the differential equation has a source term as given below:

$$\frac{d^2\xi(t)}{dt^2} + 2\beta\frac{d\xi(t)}{dt} + \omega_0^2\xi(t) = \frac{f_{dr}}{m} \tag{1.5}$$

Here f_{dr} is the driving force and m is the mass.

Solving equation of motion

First, solve the equation of motion (1.4) for unforced oscillation (See p. 1–8 of [5] or Sect. 2.7 of [6]). Set the solution in the following form:

$$\xi(t) = Ae^{\lambda t} \tag{1.6}$$

Substitution of (1.6) into differential Eq. (1.4) yields the following equation:

$$\lambda^2 Ae^{\lambda t} + \lambda 2\beta Ae^{\lambda t} + \omega_0^2 Ae^{\lambda t} = 0$$

Since $Ae^{\lambda t} \neq 0$, we can divide the above equation by $Ae^{\lambda t}$ and obtain the following equation called the characteristic equation of the differential equation.

$$\lambda^2 + 2\beta\lambda + \omega_0^2 = 0 \tag{1.7}$$

It is readily proved that $exp(\lambda t)$ satisfies differential Eq. (1.4). The two roots of Eq. (1.7) are found as follows.

$$\lambda_\pm = -\beta \pm \sqrt{\beta^2 - \omega_0^2} \tag{1.8}$$

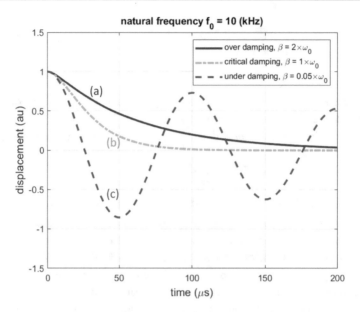

Fig. 1.2 Over-, critical-, and under-damping wave forms

Thus, we can put the general solution to Eq. (1.4) in the following form:

$$\xi(t) = A_1 e^{\left(-\beta+\sqrt{\beta^2-\omega_0^2}\right)t} + A_2 e^{\left(-\beta-\sqrt{\beta^2-\omega_0^2}\right)t} \tag{1.9}$$

The nature of solution $\xi(t)$ varies depending on the relative magnitude of β and ω_0. Generally, we classify the behavior of $\xi(t)$ into three cases called (a) over-damping, (b) critical damping, and (c) under-damping. Figure 1.2 illustrates the three cases.

As illustrated in Fig. 1.2, cases (a) and (b) do not represent an oscillatory behavior of the displacement. Since we are interested in oscillations, we discuss case (c). Below, we consider under-damped oscillation under unforced and forced conditions.

Unforced oscillation

When the decay constant β is smaller than the natural frequency ω_0, we say that the oscillation is under-damped. The discriminant in the root of the characteristic Eq. (1.7) is negative, and the characteristic equation yields complex solutions. Putting $\omega = \sqrt{\omega_0^2 - \beta^2}$ in Eq. (1.8), we obtain the following expression for the roots:

$$\lambda = -\beta \pm \sqrt{\beta^2 - \omega_0^2} = -\beta \pm i\sqrt{\omega_0^2 - \beta^2} = -\beta \pm i\omega \tag{1.10}$$

With expression (1.10), the general solution for the under-damping case becomes as follows:

$$\xi(t) = e^{-\beta t}\left(C_1 e^{i\omega t} + C_2 e^{-i\omega t}\right) = e^{-\beta t}\{(C_1 + C_2)\cos\omega t + i\,(C_1 - C_2)\sin\omega t\}$$
$$= C_0 e^{-\beta t}\cos(\omega t - \delta_0) \tag{1.11}$$

Here C_1 and C_2 are constant, and have the following relationships to the amplitude C_0 and phase δ_0.

$$C_0 = \sqrt{2(C_1^2 + C_2^2)}, \qquad \tan\delta_0 = \frac{i(C_1 - C_2)}{C_1 + C_2} \tag{1.12}$$

C_0 and δ_0 are determined with initial conditions. As an example, consider the following conditions:

$$\xi(0) = 1 \tag{1.13}$$
$$\dot{\xi}(0) = 0 \tag{1.14}$$

Condition (1.13) means that the point mass is distance 1 (in arbitrary unit) away from the equilibrium point at $t = 0$. Condition (1.14) means that the point mass is initially at rest. Using Eq. (1.11) in Eqs. (1.13) and (1.14), we obtain the following set of equations.

$$C_0 \cos(\delta_0) = 1 \tag{1.15}$$
$$-\beta\cos\delta_0 + \omega\sin\delta_0 = 0 \tag{1.16}$$

With conditions (1.15) and (1.16), we can write the general solution as follows:

$$\xi(t) = \frac{1}{\cos\delta_0}e^{-\beta t}\cos(\omega t - \delta_0) \tag{1.17}$$

where

$$\omega = \sqrt{\omega_0^2 - \beta^2} \tag{1.18}$$

$$\delta_0 = \tan^{-1}\left(\frac{\beta}{\omega}\right) \tag{1.19}$$

In the paragraph below Eqs. (1.2) and (1.3), we characterized ω_0 and β as the natural frequency and the decay constant. Here Eqs. (1.17) and (1.18) explicitly indicate the meaning of these two quantities. According to Eq. (1.17), an under-damped oscillation has the decaying feature expressed by the exponential term and the oscillatory feature expressed by the cosine term. The decay term indicates that the oscillation decays at a rate of β; the wave's amplitude decreases β in unit time. In Eq. (1.18), the (angular) frequency of the oscillatory term becomes equal to ω_0 when the decay constant is null. In other words, the natural frequency is the oscillation frequency intrinsic to the spring–mass system. It is the frequency that the system naturally oscillates when the oscillation is not disturbed by a damping force.

Forced oscillation

When an external agent drives the object connected to a spring, the resultant oscillation is called forced oscillation. When the external force is harmonic, we call the forced oscillation damped driven harmonic oscillation [7]. Many physical systems exhibit damped driven harmonic oscillation. It is a practically significant oscillation.

We can derive the equation of motion for a damped driven harmonic oscillation by using a sinusoidal function for the driving force term in Eq. (1.5).

$$\frac{d^2\xi(t)}{dt^2} + 2\beta\frac{d\xi(t)}{dt} + \omega_0^2\xi(t) = \frac{f_{ex}}{m}\sin\Omega t \tag{1.20}$$

The general solution to Eq. (1.20) consists of homogeneous and particular solutions [6]. The homogeneous solution is the solution when the right-hand side of Eq. (1.20) is zero. Physically, it represents unforced oscillation expressed by the homogeneous differential Eq. (1.4). As indicated by solution (1.17), the homogeneous solution decays over time. The particular solution is a solution that satisfies the nonhomogeneous differential Eq. (1.20) for a driving force expressed by the right-hand side. In engineering, often, the homogeneous solution represents the transient effect that dies out in a short time. On the other hand, the particular solution exhibits the steady behavior of the oscillation system and is more significant. So below, we omit the discussion on the homogeneous solution.

Consider the following test solution. In the case of unforced oscillation, the oscillatory solution (the under-damping case) has an exponential decay term multiplied in front of the sinusoidal term (1.17). You may wonder why we do not use the exponential term this time. The reason will become clear shortly in this section.

$$\xi(t) = A\sin(\Omega t - \delta) \tag{1.21}$$

Substitution of Eq. (1.21) into Eq. (1.20) yields the following equation:

$$-\Omega^2\sin(\Omega t - \delta) + 2\beta\Omega\cos(\Omega t - \delta) + \omega_0^2\sin(\Omega t - \delta) = \frac{f_{ex}}{Am}\sin\Omega t \tag{1.22}$$

By expanding $\sin(\Omega t - \delta)$ and $\cos(\Omega t - \delta)$ terms, we can rewrite Eq. (1.22) into the following form:

$$\left((\omega_0^2 - \Omega^2)\cos\delta + 2\beta\Omega\sin\delta - \frac{f_{ex}}{Am}\right)\sin(\Omega t)$$

$$+\left(-(\omega_0^2 - \Omega^2)\sin\delta + 2\beta\Omega\cos\delta\right)\cos\Omega t = 0 \tag{1.23}$$

For Eq. (1.23) to hold for all time t, it is necessary that the content of the parenthesis in front of the $\sin\Omega t$ and $\cos\Omega t$ is zero.

$$(\omega_0^2 - \Omega^2)\cos\delta + 2\beta\Omega\sin\delta = \frac{f_{ex}}{Am} \tag{1.24}$$

$$2\beta\Omega\cos\delta - (\omega_0^2 - \Omega^2)\sin\delta = 0 \tag{1.25}$$

By solving Eqs. (1.24) and (1.25) for $\cos\delta$ and $\sin\delta$, we obtain the following equations:

$$\cos\delta = \frac{f_{ex}}{Am}\frac{\omega_0^2 - \Omega^2}{(\omega_0^2 - \Omega^2)^2 + (2\beta\Omega)^2} \tag{1.26}$$

$$\sin\delta = \frac{f_{ex}}{Am}\frac{2\beta\Omega}{(\omega_0^2 - \Omega^2)^2 + (2\beta\Omega)^2} \tag{1.27}$$

It follows that the amplitude A and phase δ take the following forms:

$$A = \frac{f_{ex}/m}{\sqrt{(\omega_0^2 - \Omega^2)^2 + (2\beta\Omega)^2}} \tag{1.28}$$

$$\delta = \tan^{-1}\left(\frac{2\beta\Omega}{\omega_0^2 - \Omega^2}\right) \tag{1.29}$$

Equation (1.28) indicates that the amplitude depends on the driving frequency relative to the natural frequency and decay constant. Note that when the driving frequency is equal to the natural frequency and the decay constant (the damping coefficient) is null, the denominator of Eq. (1.28) becomes null, making the amplitude infinite. In reality, the decay constant is not null. There is always some loss mechanism that decays the oscillation.

The resonant frequency is the driving frequency that maximizes the amplitude. By finding the driving frequency that minimizes the denominator of Eq. (1.28), we find the following expression of the resonant frequency for the present case:

$$\Omega = \sqrt{\omega^2 - 2\beta^2} \tag{1.30}$$

In a system without damping, the resonant frequency is equal to the natural frequency.

1.1.2 Frequency Domain Expression

Above, we discussed exponentially decaying sinusoidal functions in the time domain. Often it is more convenient to express them in the frequency domain. In this section, we consider the Fourier transform of these functions. Later in this book, we use the frequency domain expression to discuss atomic emission spectra and optical resonator characteristics.

In the frequency domain analysis, the following formula, known as Euler's formula [8, 9], is convenient:

$$e^{i\omega t} = \cos\omega t + i\sin\omega t \tag{1.31}$$

From Eq. (1.31), we readily obtain the following expressions for the cosine and sine functions:

$$\cos \omega t = \frac{1}{2}\left(e^{i\omega t} + e^{-i\omega t}\right) \tag{1.32}$$

$$\sin \omega t = \frac{1}{2i}\left(e^{i\omega t} - e^{-i\omega t}\right) \tag{1.33}$$

Now consider the Fourier transform [10] of the cosine function. By definition, we can evaluate it as follows:

$$\mathcal{F}\left\{e^{-\beta t}\cos \omega_0 t\right\} = \int_{-\infty}^{\infty} \frac{1}{2}e^{-\beta t}\left(e^{i\omega_0 t} + e^{-i\omega_0 t}\right)e^{-i\omega t}\,dt$$

$$= \frac{1}{2}\int_0^{\infty}\left(e^{(-\beta + i(\omega_0 - \omega))t} + e^{-\beta - i(\omega_0 + \omega))t}\right)dt$$

$$= \frac{1}{2}\left\{\frac{1}{(\beta + i\omega) - i\omega_0} + \frac{1}{(\beta + i\omega) + i\omega_0)}\right\} = \frac{\beta + i\omega}{(\beta + i\omega)^2 + \omega_0^2} \tag{1.34}$$

Here, we use Eq. (1.32) in the first line and change the lower limit of the integration from $-\infty$ to 0 in the second line because we define the exponentially decaying cosine function from $t = 0$.

Similar to Eq. (1.34), using Eq. (1.33), we obtain the following expression for the Fourier transform of the decaying sine function.

$$\mathcal{F}\left\{e^{-\beta t}\sin \omega_0 t\right\} = \frac{1}{2i}\left\{\frac{1}{(\beta + i\omega) - i\omega_0} - \frac{1}{(\beta + i\omega) + i\omega_0)}\right\} = \frac{\omega_0}{(\beta + i\omega)^2 + \omega_0^2}$$

$$\tag{1.35}$$

Comparison of Eqs. (1.34) and (1.35) tells us that although the final results are different, the Fourier transforms of the decaying cosine and sine functions have the same term $1/((\beta + i\omega) - i\omega_0)$, which originates from the $exp(i\omega)$ term of Euler's formula (called the plus term). It is interesting to compare this term and the other term $1/((\beta + i\omega) + i\omega_0)$ originating from the $exp(-i\omega)$ (called the minus term). Figure 1.3 compares the absolute values of these two terms and the addition of the two terms. Notice that the minus term is significantly smaller than the plus term, which makes the plus term overlap with the addition of the two terms.

The above observation indicates that the Fourier transform of the decaying cosine and sine functions has approximately the same Fourier spectrum. This spectral shape is known as Lorentzian [11] (the Lorentz distribution also known as the Cauchy distribution [12]), which is significant in many scientific processes. Figure 1.4 shows the decaying sinusoidal function in the time and frequency domains for two decay constants with the same oscillation frequency.

Before ending this section, let's consider normalizing the Lorentzian. From Eqs. (1.34) and (1.35),

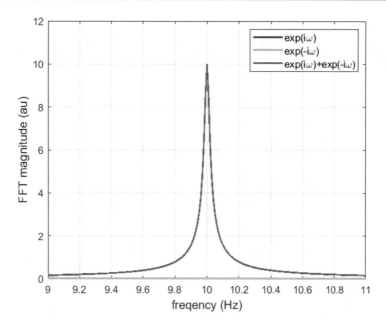

Fig. 1.3 Exponential plus term, minus term, and addition of two terms

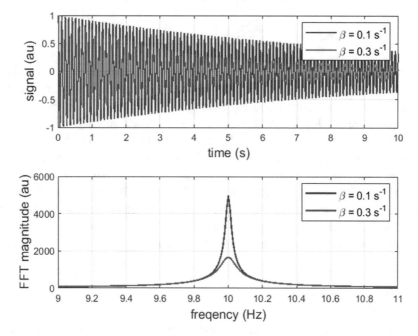

Fig. 1.4 Decaying cosine function time and frequency series

Table 1.1 Lorentzian

Function	Peak value	HWHM
$\dfrac{\beta/\pi}{\beta^2 + (\omega - \omega_0)^2}$	$\dfrac{1}{\beta\pi}$	β

$$\left| \mathcal{F}\left\{ e^{-\beta t} \cos \omega_0 t \right\} \right|^2 \cong \left| \mathcal{F}\left\{ e^{-\beta t} \sin \omega_0 t \right\} \right|^2 \cong \left| \frac{1/2}{\beta + i(\omega - \omega_0)} \right|^2$$

$$= \frac{1/4}{\beta^2 + (\omega - \omega_0)^2} \rightarrow |F(\omega)|^2 = \frac{a}{\beta^2 + (\omega - \omega_0)^2} \tag{1.36}$$

Here in the last step, we replaced $1/4$ with a to normalize this function.

$$\int_{-\infty}^{\infty} |F(\omega)|^2 d\omega = \int_{-\infty}^{\infty} \frac{a}{\beta^2 + (\omega - \omega_0)^2} d\omega = \frac{a}{\beta^2} \int_{-\infty}^{\infty} \frac{d\omega}{1 + \left(\frac{\omega - \omega_0}{\beta}\right)^2}$$

$$= \frac{a}{\beta^2} \int_{-\infty}^{\infty} \frac{\beta d\eta}{1 + \eta^2} = \frac{a}{\beta} \left[\tan^{-1} \eta \right]_{-\infty}^{\infty} = \frac{a}{\beta} \left[\frac{\pi}{2} - \left(-\frac{\pi}{2} \right) \right] = \frac{\pi a}{\beta} = 1 \tag{1.37}$$

Here $\eta = (\omega - \omega_0)/\beta$. We find the normalization factor $a = \beta/\pi$.

$$|F(\omega)|^2 = \frac{\beta/\pi}{\beta^2 + (\omega - \omega_0)^2} = \frac{1}{\beta\pi} \left[\frac{\beta^2}{\beta^2 + (\omega - \omega_0)^2} \right] \tag{1.38}$$

Note that the quantity inside the bracket on the right-hand side of Eq. (1.38) takes the maximum value of unity when the driving frequency is equal to the natural frequency, $\omega = \omega_0$. Therefore, the factor $1/(\beta\pi)$ in front of the bracket actually represents the peak (maximum) value of $|F(\omega)|^2$. Also, when $\omega - \omega_0 = \beta$, the quantity inside the bracket becomes half of the peak value $(\beta^2/(\beta^2 + \beta^2) = 1/2$. Therefore, we can characterize β as the spectral width such that at $\omega = \omega_0 \pm \beta$ the Lorentzian function is half of the peak value. The width β is called the Half Width at Half Maximum (HWHM). The width 2β is called the Full Width at Half Maximum (FWHM). Figure 1.5a illustrates sample normalized Lorentzian curves in the form of Eq. (1.38) where 2β is indicated at the half maximum height for each curve. Figure 1.5b plots the Lorentz functions in 1.5 (a) relative to the peak value. The relative height of 0.5 is observed at $\omega = \beta$ (Table 1.1).

1.1.3 Oscillation to Wave

Above we discussed harmonic oscillation using a system of a single spring and mass. With multiple springs and masses, we can extend the discussion to explain wave dynamics. First, consider a system consisting of two springs and two point masses in Fig. 1.6a. The two masses (mass A and B) are identical to each other, each $m/2$. The springs are massless. The

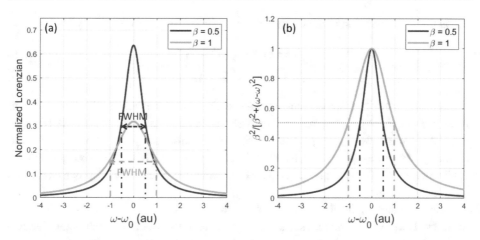

Fig. 1.5 a Normalized Lorentzian function and **b** Lorentzian function divided by peak height

left spring (Spring 1 with spring constant k_1) connects mass A to a fixed wall, and the other (Spring 2 with spring constant k_2) connects the two point masses. An external agent applies a tensile force, F, to the right end of the system. This external force causes the displacement ξ_A and ξ_B as indicated in Fig. 1.6a.

The net external force on mass A, f_A, is the vector addition of the leftward spring force exerted by Spring 1 and the rightward force by Spring 2. The net external force on mass B,

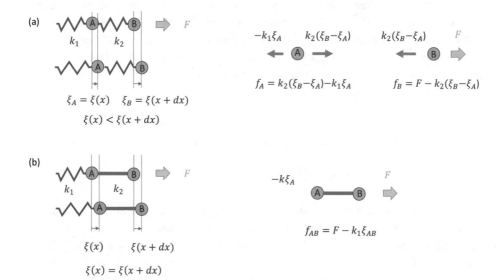

Fig. 1.6 Spring–mass model **a** k_1 and k_2 are similar to each other. **b** $k_1 \ll k_2$

f_B, is the vector addition of the rightward F and the leftward spring force by Spring 2.

$$f_A = k_2(\xi_B - \xi_A) - k_1\xi_A \tag{1.39}$$

$$f_B = F - k_2(\xi_B - \xi_A) \tag{1.40}$$

Here ξ_A and ξ_B are the displacement of the respective mass from their equilibrium positions. Note that generally $\xi_B \geq \xi_A$ because both springs stretch when F is a tensile force. In Fig. 1.6a, we express ξ_A as $\xi(x)$ to indicate that equilibrium position of mass A is x. Similarly, we express $\xi_B = \xi(x + dx)$ to indicate that mass B is at $x + dx$ prior to the application of F. These expressions are useful when we discuss waves propagating in a continuous medium.

Below, we discuss the above dynamics in two scenarios.

Scenario 1: $k_1 = k_2$ In the first scenario, the two springs have the same strength. The two masses feel the respective external forces as Eqs. (1.39) and (1.40) indicate. In this case, the external force F acts only on mass A as mass A and B are independent entities. Mass A, in turn, exerts rightward force $k_2(\xi_B - \xi_A)$ on mass B through the massless Spring 2. The displacement experienced by mass B is the addition of the displacement experienced by A and the displacement of mass B relative to mass A. In this scenario, the two springs experience the same elongation (stretch) as they have the same strength. Consequently, when the system is in equilibrium with static force F, mass B's displacement from the initial position is twice as large as mass A. Note that this does not mean that mass B feels a twice stronger force from the spring than mass A. The rightward displacement of mass A shifts the left end of Spring 2 to the right so that the two springs have the same stretch.

In the above situation, since masses are point masses, it makes more sense to interpret the force on them as stress (a force acting on a plane). (In Fig. 1.6 we interpret that two point masses are connected in series with an infinitesimally short spring.) Accordingly, it makes sense to express Hooke's law on a plane perpendicular to force F. This idea leads to the concept of strain, defined as a stretch per unit length.

$$\epsilon = \frac{\partial \xi}{\partial x} \tag{1.41}$$

Using stress and strain, we can express Hooke's law as follows:

$$\sigma = E\epsilon = E\frac{\partial \xi}{\partial x} \tag{1.42}$$

Here σ is the stress and ϵ is the strain. E is Young's modulus, the elastic constant that represents Hooke's law for a plane. Being a force on a plane σ is in N/m^2. Similarly, ϵ, stretch per unit length, is m/m and dimensionless. Consequently, the unit of Young's modulus is $(N/m^2)/[1] = N/m^2$. Here [1] denotes dimensionless. In the case of fluid media, e.g., air and water, we use pressure (N/m^2) instead of stress, and a bulk modulus for Young's modulus (see below).

Take a moment and consider the relationship between the spring constant and Young's modulus. Both represent the elasticity of a medium. The difference is that we use the former for an object with finite dimension and the latter for a plane. Hooke's law states that the elastic force is proportional to the elongation. Consequently, in static equilibrium, the longer the object, the same applied force causes it to stretch more. In other words, the spring constant is a constant for an object, not the medium. Young's modulus, on the other hand, represents the elasticity for a unit length. As long as the medium is uniform in elasticity Young's modulus is the same at all points of the object. So, Young's modulus is a medium constant.

Next, we consider the energy the external agent stores in a continuous medium. The external agent applies a tensile force at the right end of the medium. As discussed above, this force causes stress along the length of the medium. At each plane, the medium on the right applies tensile stress onto the medium on the left. As a result, the medium on the left increases its tensile strain. We can say that the mechanical work done by the medium on the right is stored as strain energy [13] in the medium on the left. The following expression represents this energy-storing process:

$$w = \int_0^{\epsilon_p} (+\sigma)(+d\epsilon) = \int_0^{\epsilon_p} (E\epsilon)d\epsilon = E \int_0^{\epsilon_p} \epsilon d\epsilon = \frac{1}{2}E\epsilon_p^2 \tag{1.43}$$

Here ϵ_p represents the maximum (peak) strain caused by σ due to the medium on the right. The plus signs in front of σ and $d\epsilon$ emphasize that when the stress due to the medium on the right is tensile ($\sigma > 0$), the resultant change in the strain is tensile ($d\epsilon > 0$).

We can make the same argument for any point of the continuous medium, saying that a portion stores strain energy onto the next portion. We can state that the work done by the external agent propagates as elastic, strain energy through the medium. We can easily imagine that this energy-storing process occurs at different portions with a time lag. If the external force acts on the right end of the medium, the part of the medium toward the left end experiences the energy-storing process later. In other words, the energy initiated by the external agent flows from the right to the left end of the medium. This energy flow constitutes an elastic compression wave.

Scenario 2: $k_1 << k_2 \to \infty$

In the second scenario, spring 2 is much stronger than spring 1, approximately infinity. In this case, the stretch of spring 2 is negligible, as Fig. 1.2b schematically illustrates. The two masses move as a combined mass, $m/2 + m/2 = m$, and $\xi_A = \xi_B$. Note that although $\xi_A - \xi_B = 0$ since k_2 is infinitely large, the product $k_2(\xi_A - \xi_B)$ has a finite value indicating that spring 2 connects the two masses. We can interpret that the force exerted by Spring 2 is the internal force for the combined system.

Consider how the external agent stores elastic energy in this case. Since Spring 2 does not stretch, the system cannot obtain strain energy. Instead, the system stores the work by the external agent as the elastic (spring) energy of Spring 1. Note that the elastic energy, in this case, is associated with the spring located outside the mass system. In a realistic situation, this spring represents the external agent's dynamics, not the medium. For instance, in the speaker system that we will consider later in this chapter, Spring 1 represents the oscillatory motion of the membrane. We can express the elastic energy as follows:

$$W_1 = \int_0^{\xi_p} (k_1\xi)d\xi = k_1 \int_0^{\xi_p} \xi d\xi = \frac{1}{2}k_1\xi_p{}^2 \tag{1.44}$$

Here W_1 is the spring potential energy stored in the combined (integrated) system, and ξ_p is the maximum displacement of the integrated system from the equilibrium point. External force F is acting on the integrated system although it has direct contact with mass A.

Scenario 3: $k_1 < k_2$

In this scenario, k_1 is the same as Scenario 1 and k_2 is greater. In this case, the system of the two point masses behaves according to Eqs. (1.39) and (1.40). The external agent stores its work as elastic energy in the point-mass system, as expressed by Eq. (1.43). As we can easily imagine, since the spring constant k_2 is greater than Scenario 1, more work is necessary to oscillate the mass system for the same peak strain. For the same peak strain, the stored energy is higher than in Scenario 1. We can express this difference with a larger Young's modulus.

In Scenario 1, we discussed that the oscillation of mass A occurs after mass B with a time lag. In Scenario 2, we found that the two masses oscillate as a single (combined) system. Being between these two cases, Scenario 3 should lead to the situation where the time lag is less than in Scenario 1. In other words, the compression wave travels faster than in Scenario 1. Indeed, the wave velocity increases with Young's modulus, as we will prove rigorously later in this chapter. Since Young's modulus is a medium constant, it follows that the wave velocity is a medium constant.

In all scenarios above, we discussed the dynamics using a tensile (pulling) external force. We can apply the same discussion to the case where an external agent applies compressive (pushing) force to the system from its left end and derive the same force Equations (1.39) and (1.40), and energy expressions (1.44) and (1.43). In the next section, as a realistic example of such a case, we discuss the dynamics of air molecules when a speaker generates a sound wave.

1.1.4   Sound Wave

Consider how a speaker [14] generates a sound wave in Fig. 1.7. The mechanical oscillator applies a sinusoidal force on the membrane. As a result, the membrane undergoes a sinusoidal displacement generating a compression wave in the air next to it. Figure 1.7a illustrates the situation where the membrane is fully extended to the right turning point compressing the air immediately next to it at the maximum pressure. As the membrane retreats leftward, the maximum pressure moves to the right because the air immediately next to the membrane compresses the air next to it. Set a x-axis in the direction of this movement with the origin

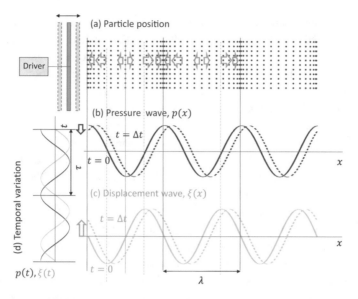

Fig. 1.7 Speaker membrane generates compression wave. τ and λ are the period and wavelength of the wave

at the right turning point of the membrane. The rest of the air shows the same pattern, i.e., the compressive wave travels in the positive x-direction.

For a one-dimensional model as shown in Fig. 1.7, we can consider the air next to the speaker as many point masses connected in series with springs of the same strength. Here, the elastic constant corresponding to the spring constant in the point-mass model is bulk modulus B [15]. We can express Hooke's law as follows:

$$p = -B\frac{\Delta V}{V} = -B\frac{S\Delta l}{Sl} = -B\frac{\partial \xi}{\partial x} \tag{1.45}$$

Here V is the volume of a block of air, ΔV is its change due to the compressive force exerted by the neighboring block of air, S is its cross-sectional area, and l and Δl are the lengths and its change of the block associated with the compression. The rightmost-hand side of Eq. (1.45) expresses the volume change rate in the infinitesimal limit. Equation (1.45) is equivalent to Eq. (1.42) except for the negative sign in front of the elastic constant (modulus). This difference comes from the definition of stress and pressure. In continuum mechanics [17–20], tensile stress is conventionally defined as positive. (The community of high-speed deformation tests [21], such as split-Hopkinson bar tests, defines compressive stress as positive.) On the other hand, a positive pressure of a fluid is compressive. In other words, when $\partial \xi/\partial x < 0$ (compressive), the pressure $-B\partial \xi/\partial x$ is positive with the bulk modulus itself positive, $B > 0$.

Figure 1.7b plots the pressure as a function of x. Here the solid line presents the pressure at the moment when the membrane is at $x = 0$ (define this moment as $t = 0$ s). The dashed line presents the pressure pattern at Δt later, explicitly exhibiting the rightward movement of the pressure wave. Note that this traveling wave represents the motion of the pressure pattern, not the air molecules'. As the membrane moves back and forth, the molecules near $x = 0$ oscillate, and the oscillation propagates in the x-direction. The vertical arrow on the left of Fig. 1.7b indicates that the pressure decreases at $x = 0$ as the pressure wave travels.

Consider now the motion of air molecules in more detail. Figure 1.7c shows the displacement of the air molecules as a function of x. Here, the solid line represents the displacement wave at $t = 0$, and the dashed line represents the situation at $t = \Delta t$. These lines indicate that the displacement wave travels to the right at the same phase velocity as the pressure wave. The vertical arrow on the left of Fig. 1.7c indicates that the molecules at $x = 0$ move to the right (in the positive x-direction) in the duration from $t = 0$ to $t = \Delta t$.

The displacement of air molecules as a function of time may not be pictured straightforwardly. We will discuss the relation between the pressure wave and displacement wave quantitatively later in this chapter. For now, the following qualitative explanation will help us understand the dynamics. The maximum pressure occurs when the molecules on the left move fully rightward and those on the right move fully leftward. The horizontal arrows inserted in Fig. 1.7a illustrate the situation. When the membrane starts moving to the left from its right turning point, the molecules near $x = 0$ enter the next phase in which the pressure reduces from the maximum. This pressure reduction accompanies the following change in the air molecules' displacement. The molecules on the left switch their directions; they start to move leftward. Similarly, the molecules on the right start to move rightward.

Acoustic energy

With Eqs. (1.42) and (1.43) we considered the strain energy stored in an elastic medium. We can discuss the acoustic (sound) energy [22] stored in the air similarly. Using Eq. (1.45) and considering Eq. (1.43), we obtain the following expression:

$$w_{ac} = \int_0^{\epsilon_p} p(-d\epsilon) = \int_0^{\epsilon_p} (-B\epsilon)(-d\epsilon) = B \int_0^{\epsilon_p} \epsilon d\epsilon = \frac{1}{2} B \epsilon_p^2 \tag{1.46}$$

Here the negative sign for $d\epsilon$ indicates that when the pressure is positive (compressive), it makes the air more compressive, meaning that the change in strain along the x-axis is negative, i.e., $d\epsilon = (\partial\epsilon/\partial x)dx < 0$.

Compression wave as energy flow

Above, we discussed that the wave velocity increases with elastic modulus. Here we briefly consider the relationship between the wave velocity and elastic modulus in a qualitative fashion. Later, we discuss it quantitatively.

As clear from the discussions with the speaker example and the point-mass system analogy, a compression wave is a motion of strain energy associated with the local oscillation of the medium. When the compression is maximum, the pressure is maximum. Figure 1.7 indicates that at the location where the pressure takes either a positive or negative peak, the molecular displacement is zero. In a harmonic oscillation of a mass, zero displacements coincide with a peak velocity. These arguments lead to the statement that at the locations where the pressure is at a peak in Fig. 1.7, the air molecules are at a peak velocity. In other words, the pressure and velocity waves take peak values simultaneously.

From the above discussion, we can envision that the pressure and velocity waves pair to carry the sound energy. Along this line of argument, we can derive the concept of sound wave velocity as the flow of strain or kinetic energy of air molecules. In Eq. (1.46), we expressed the acoustic strain energy as $(B/2)\epsilon^2$. We can identify the kinetic energy of the air molecule as $(\rho/2)v^2$. Here, ρ is the density of air.

In the above argument, both forms of energy originate from the elastic oscillatory dynamics of the neighboring segment of the air, in which total kinetic energy is the same as total potential energy. In the process of the air molecules' oscillation, the kinetic and the spring potential energies alternate. When the molecules are at equilibrium, the kinetic energy is at maximum with zero potential energy; when they are at turning points, the potential energy is at maximum with null kinetic energy. The acoustic strain energy $(B/2)\epsilon^2$ is due to the potential energy, and the kinetic energy $(\rho/2)v^2$ is due to the air molecules' kinetic energy. Thus, we can equate them as follows:

$$\frac{E}{2}\epsilon^2 = \frac{\rho}{2}v^2 \tag{1.47}$$

Considering that the strain is the displacement per unit length, $\epsilon = \Delta\xi/\Delta x$ and the velocity is the displacement per unit time, $v = \Delta\xi/\Delta t$, we can rewrite Eq. (1.47) into the following form:

$$\frac{v}{\epsilon} = \frac{\Delta\xi}{\Delta t} / \frac{\Delta\xi}{\Delta x} = \frac{\Delta x}{\Delta t} = \sqrt{\frac{E}{\rho}} \tag{1.48}$$

The meaning of Δx and Δt in Eq. (1.48) needs some attention. Δx represents the infinitesimal distance we use to define strain. Similarly, Δt represents the time to define velocity. In wave dynamics, a spatial change is meaningful in association with spatial periodicity, i.e., the wavelength λ. If the wavelength is longer, we can take a larger Δx to define strain. A temporal change is meaningful in association with temporal periodicity, i.e., the period τ. By shifting along the spatial axis by a certain amount, we observe the corresponding change in the wave quantity. For instance, if we shift by half wavelength, a sinusoidal wave changes its sign. Similarly, by shifting by half period along the time axis, the wave changes the sign. If we normalize the spatial and temporal shift by the respective periodicity, the associated change in the wave is the same.

$$\frac{\Delta x}{\lambda} = \frac{\Delta t}{\tau} \tag{1.49}$$

Equation (1.49) leads to the following equation:

$$\frac{\Delta x}{\Delta t} = \frac{\lambda}{\tau} \equiv v_{ph} \tag{1.50}$$

The quantity λ/τ is known as the phase velocity of a wave, v_{ph}. It represents the ratio of the length of one period of a wave counted in the space axis over to the time axis. Take a look at Fig. 1.7c and d. If we measure one period of the displacement wave, it is λ (m). If we measure it in time, it is τ (s). The wave velocity represents the ratio of the former to the latter. Shortly we will find out why this ratio indicates the velocity of the wave.

From Eqs. (1.48) and (1.50), we find the quantity $\sqrt{E/\rho}$ represents the phase velocity of the strain wave.

$$\sqrt{\frac{E}{\rho}} = \frac{\Delta x}{\Delta t} = \frac{\lambda}{\tau} = v_{ph} \tag{1.51}$$

This finding is consistent with the above qualitative argument that the wave velocity should increase with Young's modulus. Shortly, we will prove that Eq. (1.51) represents the elastic compression wave velocity by solving a wave equation.

Here we discussed a compression wave in solids using Young's modulus. The same discussion holds for an acoustic wave by replacing Young's modulus with bulk modulus.

Acoustic energy wave

The above paragraph tells us an acoustic wave consists of stress (pressure) and velocity waves. Naturally, we can envision that their product represents the compression (acoustic) energy flow. Consider the units of this product. Stress is in N/m^2 and velocity is in m/s. Therefore, the unit of the product is (N·m)/(m^2·s) = J/(m^2·s)=(J/m^3)·(m/s). We can interpret this product as the energy density represented by (J/m^3) flows with the velocity represented by (m/s). Further, we can interpret that the energy density is a combination of stress energy and particles' kinetic energy. More specifically, the following argument holds. Expressing the stress and particle velocity as $\sigma(t)$ and $v(t)$,

$$< \epsilon^2 >_a vg = \sigma v = E\epsilon v = E\epsilon \left(\sqrt{\frac{E}{\rho}}\epsilon\right) = E\epsilon^2 \sqrt{\frac{E}{\rho}} = E\epsilon^2 v_{ph} \tag{1.52}$$

$$< v^2 >_a vg = \sigma v = E\epsilon v = E\left(\sqrt{\frac{\rho}{E}}v\right)v = \rho v^2 \sqrt{\frac{E}{\rho}} = \rho v^2 v_{ph} \tag{1.53}$$

Here we used Eq. (1.48) in going through the second equal sign and Eq. (1.51) in going through the last equal sign.

Expressing the stress and velocity as $\epsilon(t) = \epsilon_p \sin \omega t$ and $v(t) = v_p \cos \omega t$, we can express the average values over one period as follows:

$$\frac{\omega}{2\pi} \int_0^{\frac{2\pi}{\omega}} \epsilon_p^2 \sin^2 \omega t \, dt = \frac{1}{2} \epsilon_p^2 \tag{1.54}$$

$$\frac{\omega}{2\pi} \int_0^{\frac{2\pi}{\omega}} v_p^2 \cos^2 \omega t \, dt = \frac{1}{2} v_p^2 \tag{1.55}$$

Thus, the average energy density flow becomes as follows:

$$(\sigma v)_{av} = \left(\frac{E}{2} \epsilon_p^2 + \frac{\rho}{2} v_p^2 \right) v_{ph} \tag{1.56}$$

Equation (1.56) indicates that the average energy density consists of the elastic potential energy and particles' kinetic energy.

1.2 Wave Equation and Solution

1.2.1 Compressive Wave Equation and Solution

Figure 1.8 illustrates a portion of a long elastic medium. An external agent applies a tensile force on the left and right ends of the medium. Consider the plane at x and $x + dx$, and call the block on the left of the plane at x the left block, the one on the right of the plane at $x + \Delta x$ the right block, and the block between them the middle block. On the plane at x, the left block exerts leftward normal stress onto the middle block. On the plane at $x + \Delta x$, the right block exerts rightward normal stress onto the middle block. We can express the stresses on these planes as follows.

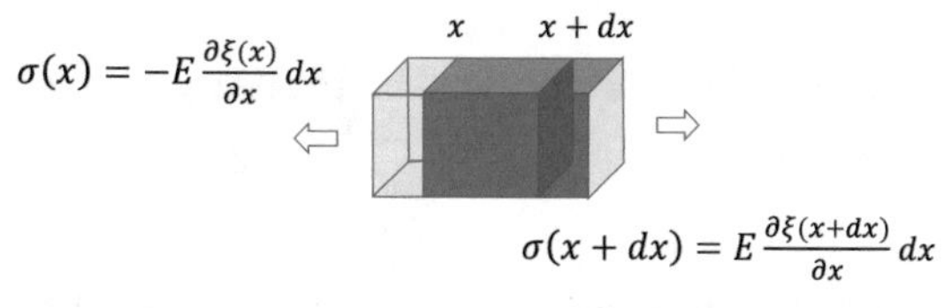

Fig. 1.8 Normal stress at x and $x + \Delta x$ in an elastic medium

$$\sigma(x) = -E \frac{\partial \xi(x)}{\partial x} dx$$

$$\sigma(x + dx) = E \frac{\partial \xi(x+dx)}{\partial x} dx$$

$$\sigma(x) = E\epsilon(x) = E\frac{\partial \xi}{\partial x}\Big|_x \tag{1.57}$$

$$\sigma(x+dx) = E\epsilon(x+dx) = E\frac{\partial \xi}{\partial x}\Big|_{x+dx} \tag{1.58}$$

Thus, the net force acting on the middle block becomes

$$S\left(\sigma(x+dx) - \sigma(x)\right) = SE\left(\frac{\partial \xi}{\partial x}\Big|_{x+dx} - \frac{\partial \xi}{\partial x}\Big|_x\right) = SE\frac{\partial^2 \xi}{\partial x^2}dx \tag{1.59}$$

Here S is the cross-sectional area. Since the mass of the medium between x and $x + dx$ is $\rho S dx$, we can write the equation of motion as follows:

$$\rho S dx\frac{\partial^2 \xi}{\partial t^2} = SE\frac{\partial^2 \xi}{\partial x^2}dx \tag{1.60}$$

Dividing the above equation by $\rho S dx$, we obtain the following differential equation:

$$\frac{\partial^2 \xi}{\partial t^2} = \frac{E}{\rho}\frac{\partial^2 \xi}{\partial x^2} \tag{1.61}$$

We can solve differential Eq. (1.61) using the variable separation. Put the solution as the product of a time function $T(t)$ and space function $X(x)$ as $\xi(t, x) = T(t)X(x)$, substitute it into the differential equation, and divide it by TX setting the result equal to a constant $-c^2$.

$$\frac{\ddot{T}}{T} = \frac{E}{\rho}\frac{X''}{X} = -c^2 \tag{1.62}$$

Equation (1.62) allows us to put the solutions $T(t)$ and $X(x)$ in the following forms.

$$T(t) = T_{01}e^{ict} + T_{02}e^{-ict} = T_c \cos(ct) + T_s \sin(ct) \tag{1.63}$$

$$X(x) = X_{01}e^{ic\sqrt{\frac{\rho}{E}}x} + X_{02}e^{-ic\sqrt{\frac{\rho}{E}}x} = X_c \cos\left(c\sqrt{\frac{\rho}{E}}x\right) + X_s \sin\left(c\sqrt{\frac{\rho}{E}}x\right) \tag{1.64}$$

Use the following boundary conditions for $X(x)$ to assimilate to Fig. 1.7.

$$X(0) = X(\lambda) = 0 \tag{1.65}$$

Here λ is the wavelength. The boundary condition (1.65) allows us to put $X(x)$ in the following form:

$$X(x) = X_s \sin\left(\frac{n\pi}{\lambda}x\right) \tag{1.66}$$

where n is an integer. Comparing Eqs. (1.64) and (1.66), we can express c as follows:

$$c = \sqrt{\frac{E}{\rho}}\frac{n\pi}{\lambda} \tag{1.67}$$

Substituting Eq. (1.67) into Eq. (1.63), we obtain the following form for the time function $T(t)$:

$$T(t) = T_c \cos\left(\sqrt{\frac{E}{\rho}}\frac{n\pi}{\lambda}t\right) + T_s \sin\left(\sqrt{\frac{E}{\rho}}\frac{n\pi}{\lambda}t\right) \tag{1.68}$$

Use the following boundary conditions for $T(t)$:

$$T(0) = T(\tau) = 0. \tag{1.69}$$

Here τ is the period. It follows that $T_c = 0$. From the conditions $T_c = 0$ and $T(\tau) = 0$, we find that the argument of the sine functions in Eq. (1.68) must be equal to $n\pi$ when $t = \tau$. This leads to the following equality:

$$\sqrt{\frac{E}{\rho}} = \frac{\lambda}{\tau} = v_{ph} \tag{1.70}$$

As we discussed with Eq. (1.50), the right-hand side of Eq. (1.70) represents the phase velocity, v_{ph}. Hence, this equation indicates that the phase velocity of this compression wave is $\sqrt{E/\rho}$.

From the first equality expressed by Eq. (1.70), we can rewrite Eq. (1.68) as follows:

$$T(t) = T_s \sin\left(\frac{n\pi}{\tau}t\right) \tag{1.71}$$

Combining expression (1.66) and (1.71), we obtain the following expression for the displacement wave as a function of time and space.

$$\xi(t, x) = (X_s T_s) \sin\left(\frac{n\pi}{\tau}t\right) \sin\left(\frac{n\pi}{\lambda}x\right) \tag{1.72}$$

$$= \frac{A}{2} \sin\left(\frac{n\omega}{2}t\right) \sin\left(\frac{nk}{2}x\right) \tag{1.73}$$

Here $A/2 = (X_s T_s)$ is the amplitude of the ξ wave, $\omega = 2\pi/\tau$ is the temporal frequency (the reciprocal of the period) measured in the unit of phase in radian, and $k = 2\pi/\lambda$ is the spatial frequency in radian.

Figure 1.9 is an example of a wave expressed in the form of Eq. (1.72). As this figure indicates, the wave does not move to the left or right with time. These types of waves are called standing waves [23].

In Eq. (1.72) v_{ph} represents the phase velocity of this wave. You may wonder why a standing wave has a velocity. We will clarify this question in the next section.

1.2.2 Standing Wave and Traveling Wave

In the preceding section, we obtained a solution to the wave Eq. (1.61) in the form of the product of the time-dependent and space-dependent functions. Note that this is a natural

Fig. 1.9 An example of
standing wave. For clarity, the
first half cycle of a period is
shown in solid line and the
second is in dashed line

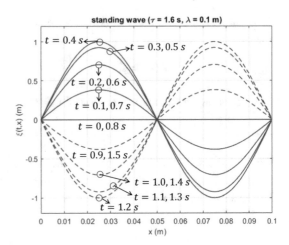

consequence of using the variable separation to solve the wave equation. In other words, we
forced the solution to take this form. In this section, we consider other forms of waves that
satisfy the same wave equation.

Using the mathematical identity, we can write Eq. (1.72) as follows:

$$\xi(t, x) = \frac{A}{2} \sin\left(\frac{n\omega}{2}t\right) \sin\left(\frac{nk}{2}x\right) = A\left(\cos(n\omega t - nkx) - \cos(n\omega t + nkx)\right) \quad (1.74)$$

We can easily see that each term of Eq. (1.74) satisfies the wave Eq. (1.61. Thus, we can say
that the wave Eq. (1.61) yields the following three types of wave solutions:

$$\xi_0(t, x) = \frac{A}{2} \sin\left(\frac{n\omega}{2}t\right) \sin\left(\frac{nk}{2}x\right) \quad\quad\quad (1.75)$$

$$\xi_1(t, x) = A_1 \cos(n\omega t - kx + \phi_1) \quad\quad\quad (1.76)$$

$$\xi_2(t, x) = A_2 \cos(n\omega t + kx + \phi_2) \quad\quad\quad (1.77)$$

Here, A_1 and A_2 are constant amplitude, and ϕ_1 and ϕ_2 are constant phases. Being constants,
these parameters do not affect the differentiations in the wave Eq. (1.61). With constant
amplitude, the wave would have the same strength over the plane perpendicular to the x-
axis at any point on the axis. Since a wave carries energy, this condition means that the
energy is infinite (as the wave covers an infinitely large area). Such a wave is unrealistic,
but we often use this type of solution as it exhibits important characteristics of the wave. We
call this type of wave a plane wave. We call the constant phase ϕ_1 and ϕ_2 the initial phase of
the respective waves, as they are independent of time (hence represent the phase at $t = 0$).
Note that $\phi_{1,2} = \pm\pi/2$ will change the cosine functions to sine functions, indicating that a
solution in the form of $A \sin(\omega \pm kx + \phi)$ also satisfies the wave Eq. (1.61). The arbitrary

integer n makes the frequency n-fold higher. We call the solution with $n > 1$ harmonics of the $n = 1$ case, which is referred to as the wave with the fundamental frequency.

Continue the discussion with solution (1.76) using $n = 1$ and $\phi_1 = 0$. These conditions do not affect the generality of the discussion made here.

$$\xi_1(t, x) = A_1 \cos(\omega t - kx) \tag{1.78}$$

Substitute solution (1.78) into Eq. (1.61).

$$A_1 \omega^2(-\cos(\omega t - kx)) = A_1 \frac{E}{\rho} k^2(-\cos(\omega t - kx)) \tag{1.79}$$

Equation (1.79) holds with the following condition:

$$\omega^2 = \frac{E}{\rho} k^2 \tag{1.80}$$

Condition (1.80) is true as long as the following equation is true:

$$\frac{E}{\rho} = \frac{\omega^2}{k^2} = \left(\frac{2\pi}{\tau} \bigg/ \frac{2\pi}{\lambda}\right)^2 = \left(\frac{\lambda}{\tau}\right)^2 \tag{1.81}$$

Equation (1.81) is equivalent to Eq. (1.70). Remembering that λ/τ is the phase velocity, we now see that $\sqrt{E/\rho}$ is the phase velocity derived from the wave Eq. (1.61).

Now consider the argument of the cosine function in Eq. (1.78). Since it is an argument of a cosine function, we call it the phase of a wave, θ.

$$\theta(t, x) = \omega t - kx \tag{1.82}$$

Using expression (1.82), consider the relation between the time variable t and space variable x for a constant phase. Since the value of a cosine function is solely dependent on the phase, a constant phase means a constant value of the wave. If the wave represents air compression, the trajectory of a constant phase is the trajectory of a certain level of compression. For example, pay attention to a constant phase $\theta = 0$. Since a cosine function takes the maximum value of unity at $\theta = 0$, the corresponding trajectory is the trace of a crest of the wave. Under this condition, increase t by Δt and find the condition of Δx that makes $\theta(t + \Delta t, x + \Delta x) = 0$.

From Eq. (1.82) and using the condition $\theta(t, x) = 0$, we obtain the following equation:

$$\theta(t + \Delta t, x + \Delta x) = \omega \cdot (t + \Delta t) - k \cdot (x + \Delta x) = (\omega t - kx) + (\omega \Delta t - k\Delta x)$$
$$= \theta(t, x) + (\omega \Delta t - k\Delta x) = 0 + \omega \Delta t - k\Delta x = 0 \tag{1.83}$$

It follows that

$$\frac{\Delta x}{\Delta t} = \frac{\omega}{k} = \frac{\lambda}{\tau} \tag{1.84}$$

Equation (1.84) is the same as Eq. (1.50), indicating that the phase velocity of a wave is the velocity of a constant phase. Sometimes we call a crest of a wave the wavefront. We can say that the phase velocity is the velocity of the wavefront.

We can extend the above discussion from a crest to any constant phase.

$$d\theta = \frac{\partial \theta}{\partial t} dt + \frac{\partial \theta}{\partial x} dx = \omega dt - k dx \tag{1.85}$$

Therefore, we can write the condition of a constant phase as follows:

$$\frac{d\theta}{dt} = \omega - k \frac{dx}{dt} = 0 \tag{1.86}$$

Equation (1.86) leads to the following expression of phase velocity:

$$v_{ph} = \frac{\omega}{k} = \frac{\lambda}{\tau} = \frac{dx}{dt} \tag{1.87}$$

Equation (1.87) is the infinitesimal form of (1.50), indicating that we can define the phase velocity from the instantaneous change in the space and time coordinate variables as we trace the wavefront.

So, now we have convinced ourselves that the phase velocity ω/k represents the motion of a wave $\xi_1(t, x)$ (1.78). Note that Eq. (1.87) indicates $v_{ph} > 0$ (because $\omega > 0$ and $k > 0$). The positive phase velocity v_{ph} means that the wave travels in the positive x-direction. We call this type of wave a traveling wave propagating in the positive x-direction.

Above, we discussed the traveling wave represented by the first term of Eq. (1.74). What about the second term? The only difference in the above procedure is that $\partial \theta / \partial x = \omega/(-k) < 0$, leading to a negative phase velocity.

$$v_{ph} = -\frac{\omega}{k} \tag{1.88}$$

The negative phase velocity indicates that the second term of Eq. (1.74) represents a traveling wave propagating in the negative x-direction.

Since both terms satisfy the wave Eq. (1.61) independently of each other, we can say that Eq. (1.61) yields traveling waves in either direction as long as the phase velocity is $\sqrt{(E/\rho)}$, in addition to the standing wave we already discussed. Either traveling in the positive or negative x-direction, the wave travels in line with the direction of particle oscillation. This type of wave is called a longitudinal wave.

A natural question would be the relation between the traveling and standing waves resulting from the same wave equation. It is clear from the above discussion that the addition of two oppositely traveling waves generates a standing wave. (See p. 144 of [23].) The mathematical identity used in Eq. (1.74) indicates that only if the coefficient for the two terms on the right-hand side of (1.74) is the same, the left-hand side takes the form of the product of the time and space functions. In terms of physics, this observation leads to the following statement: if oppositely propagating traveling waves at the same phase velocity have the same amplitude, they form a standing wave. If one has greater amplitude, the summed wave has a traveling wave component. This observation is closely related to the concept known as the resonance of waves and laser resonators. We will discuss these concepts later in this chapter.

1.2.3   Three-Dimensional Compression Wave Equation

Above, we discussed the compression wave in one dimension (Eq. (1.61)). We can extend the same argument into three dimensions. By replacing the one-dimensional spatial differentiation $\partial^2/\partial x^2$ with Laplacian ∇^2 [24], we can write the three-dimensional version of Eq. (1.61) in the following form [25]:

$$\frac{\partial^2 \boldsymbol{\xi}}{\partial t^2} = \frac{\lambda + 2G}{\rho} \nabla^2 \boldsymbol{\xi} \tag{1.89}$$

$$v_p = \sqrt{\frac{\lambda + 2G}{\rho}} \tag{1.90}$$

Here λ and G are the first and second Lamé parameters [19]. We can view the quantity $\lambda + 2G$ as the three-dimensional version of Young's modulus. Equation (1.90) is the phase velocity of the three-dimensional wave, corresponding to the one-dimensional case expressed by Eq. (1.70).

It is worthwhile considering the three-dimensional wave equation a little deeper. Using the mathematical identity $\nabla^2 \mathbf{A} = \nabla(\nabla \cdot \mathbf{A}) - \nabla \times (\nabla \times \mathbf{A})$ ($\mathbf{A}$ is an arbitrary vector), we can rewrite Eq. (1.89) as follows:

$$\frac{\partial^2 \boldsymbol{\xi}}{\partial t^2} = \frac{\lambda + 2G}{\rho} \left(\nabla(\nabla \cdot \boldsymbol{\xi}) - \nabla \times (\nabla \times \boldsymbol{\xi}) \right) \tag{1.91}$$

Now take the divergence of Eq. (1.91) and use the identity $\nabla \cdot (\nabla \times \mathbf{A}) = 0$ and $\nabla \cdot \nabla = \nabla^2$. We obtain the following equation:

$$\frac{\partial^2 (\nabla \cdot \boldsymbol{\xi})}{\partial t^2} = \frac{\lambda + 2G}{\rho} \nabla^2 (\nabla \cdot \boldsymbol{\xi}) \tag{1.92}$$

The quantity $\nabla \cdot \boldsymbol{\xi}$ is known as the volume expansion in continuum mechanics [25]. Equation (1.92) represents the wave equation of volume expansion (rarefaction). In Cartesian coordinates, we can express the volume expansion as follows:

$$(\nabla \cdot \boldsymbol{\xi}) = \left(\frac{\partial \xi_x}{\partial x} + \frac{\partial \xi_y}{\partial y} + \frac{\partial \xi_z}{\partial z} \right) \tag{1.93}$$

According to Poisson's effect, all the derivative terms on the right-hand side of Eq. (1.93) cannot take the same sign. If the volume expands in a certain direction, it compresses in the orthogonal directions. Equation (1.92) indicates that the elastic medium takes this pattern of deformation and the pattern propagates as a wave.

1.2.4 Transverse Mechanical Wave

So far, we have paid attention to compression (longitudinal) wave dynamics based on linear Hooke's law. Electromagnetic waves are transverse waves, i.e., the direction of particle oscillation is orthogonal to the direction of wave propagation. It is worthwhile discussing transverse mechanical wave dynamics before considering electromagnetic wave dynamics.

In mechanical wave dynamics, a transverse wave is generated by shear force. Consider vertical shear force in two dimensions in Fig. 1.10. In this figure an elastic block whose surface is in the xy plane experiences shear strain $\epsilon_s(x) = \partial \eta(x)/\partial x$. Here $\eta(x)$ denotes the vertical displacement at x. Assume that the vertical displacement is upward ($\eta(x) > 0$) and the shear strain increases with x ($\partial \epsilon_x(x)/\partial x = \partial^2 \eta(x)/\partial x^2 > 0$). Under this condition, consider three segments in this block; the left, middle, and right blocks. The left and middle blocks share a plane at x, and the middle and right blocks share a plane at $x + \Delta x$. At the boundary at x, the left face of the middle block is displaced more upward than the right face of the left block (because partial $\eta/\partial x > 0$). Consequently, due to the elasticity of the material, the left face of the middle block is pulled downward by the left block. By the same mechanism, at the boundary at $x + \Delta x$, the left face of the right block is pulled downward by the middle block. From the perspective of the middle block, its right face is pulled upward by the right block (the reaction to the force that the middle block exerts on the right block). Thus, the middle block is pulled upward at the $x + \Delta x$ end face and downward at the x end.

The above analysis leads to the following equation of motion for the middle block:

$$\rho S dx \frac{d^2\eta}{dt^2} = G \left(\frac{\partial \eta}{\partial x}\bigg|_{x+\Delta x} - \frac{\partial \eta}{\partial x}\bigg|_{x} \right) = G \left(\frac{\partial \eta(x+dx)}{dx} - \frac{\partial \eta(x)}{dx} \right) S = GS\frac{\partial^2\eta}{\partial x^2} S dx$$

(1.94)

Equation (1.94) leads to the following one-dimensional transverse wave equation:

$$\frac{\partial^2\eta}{\partial t^2} = \frac{G}{\rho}\frac{\partial^2\eta}{\partial x^2}$$

(1.95)

Fig. 1.10 Shear force generating a shear wave

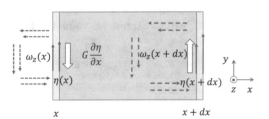

Noting that Eq. (1.95) has the same form as Eq. (1.61), we can easily see that this wave equation yields traveling or standing wave solutions of the same type as Eqs. (1.75), (1.76), and (1.77)

$$\eta_0(t, x) = \frac{A}{2} \sin\left(\frac{n\omega}{2}t\right) \sin\left(\frac{nk}{2}x\right) \tag{1.96}$$

$$\eta_1(t, x) = A_1 \cos(n\omega t - kx + \phi_1) \tag{1.97}$$

$$\eta_2(t, x) = A_2 \cos(n\omega t + kx + \phi_2) \tag{1.98}$$

These wave solutions are transverse waves, representing the vertical (y) displacement's horizontal (x) dependence, i.e., the wave travels parallel to the x-axis whereas the oscillation is parallel to the y-axis.

By viewing η as the y-component of displacement vector $\boldsymbol{\xi}$, we can interpret $\partial\eta/\partial x$ as part of the z component of $\boldsymbol{\omega} \equiv \nabla \times \boldsymbol{\xi}$.

$$\omega_z = (\nabla \times \boldsymbol{\xi})_z = \left(\frac{\partial\eta}{\partial x} - \frac{\partial\xi}{\partial y}\right) \tag{1.99}$$

Further, we can interpret the differential shear stress (the quantity inside the parenthesis on the middle term of Eq. (1.94)) as $\partial\omega_z/\partial x$, which is part of the y-component of vector $\nabla \times \boldsymbol{\omega}$. With this interpretation, we can write the equation of motion as follows:

$$\rho\frac{\partial^2\eta}{\partial t^2} = -G\,(\nabla \times \boldsymbol{\omega})_y = -G\left(\frac{\partial\omega_x}{\partial z} - \frac{\partial\omega_z}{\partial x}\right) \tag{1.100}$$

Ignoring the x-components of $\boldsymbol{\xi}$ and $\boldsymbol{\omega}$ and substituting Eq. (1.99) into the second term, we can rewrite Eq. (1.100) into the following form:

$$\rho\frac{\partial^2\eta}{\partial t^2} = -G\frac{\partial}{\partial x}\left(-\frac{\partial\omega_z}{\partial x}\right) = G\frac{\partial^2\eta}{\partial x^2} \tag{1.101}$$

Equation (1.101) is equivalent to Eq. (1.95). In other words, we can interpret the one-dimensional equation of motion (1.95) as part of the y-component of the following set of equations:

$$\nabla \times \boldsymbol{\xi} = \boldsymbol{\omega} \tag{1.102}$$

$$\nabla \times \boldsymbol{\omega} = -\frac{\rho}{G}\frac{\partial^2\boldsymbol{\xi}}{\partial t^2} \tag{1.103}$$

As we will see shortly, the set of Eqs. (1.102) and (1.103) is analogous to Maxwell's equation, which governs the wave dynamics of the electromagnetic field.

1.3   Resonance

Resonance is a phenomenon in which an external agent transfers energy to an oscillatory system with the highest efficiency. Under a resonant condition, the oscillatory system exhibits the maximum amplitude. In this section, we discuss resonance observed in harmonic oscillations and waves.

1.3.1   Resonance in Harmonic Oscillations

In harmonic oscillations, resonance represents the situation where an oscillatory system driven by an external agent shows significantly higher amplitude at a driving frequency than other frequencies. In the paragraph under Eq. (1.29), we discussed that the amplitude of a damped driven harmonic oscillator depends on the driving frequency. By finding the driving frequency that minimizes the denominator of the amplitude expression (1.28), we find the resonant (angular) frequency for a system whose natural frequency is ω_0, and the decay constant is β as given below:

$$\Omega = \sqrt{\omega^2 - 2\beta^2} \tag{1.104}$$

Equation (1.104) indicates that in an oscillatory system without a damping mechanism ($\beta = 0$), the natural frequency is the resonant frequency. In this case, the resonant frequency is equal to $\sqrt{k_{sp}/m}$ (Eq. (1.2)). This statement leads to the following observation regarding the phase velocity.

Remembering that Young's modulus is the spring force per unit area, we obtain the following expressions:

$$k_{sp}dx = ES \tag{1.105}$$

$$E = \frac{k_{sp}dx}{S} \tag{1.106}$$

Here S is the cross-sectional area perpendicular to dx.

The phase velocity of an elastic medium having Young's modulus of E and density ρ is $\sqrt{E/\rho}$ (Eq. (1.48)). Using Eq. (1.106) we can write this phase velocity expression as follows:

$$v_p = \sqrt{\frac{E}{\rho}} = \sqrt{\frac{k_{sp}dx}{S\rho}} = \sqrt{\frac{k_{sp}dx}{S(m/(Sdx))}} = \sqrt{\frac{k_{sp}}{m}}dx = \omega_0\frac{\lambda}{2\pi} = v_0\lambda \tag{1.107}$$

Here $\sqrt{k_{sp}/m}$ is the angular resonant frequency of mass m connected to a spring of spring constant k_{sp} (see the paragraph under Eq. (1.3) where we discussed that this quantity is the natural frequency of the spring–mass system), and dx is the spatial span of the mass. By interpreting dx as the spatial periodicity $dx = 1/k = \lambda/(2\pi)$, we can view Eq. (1.107) as taking the form of "frequency × wavelength" ($\omega_0 = 2\pi \nu_0$, where ω_0 and ν_0 are the resonant angular frequency and frequency). In this view, we can interpret the phase velocity as originating from the propagation of the resonant oscillation of the unit mass. In other words, the compression wave is the propagation of the medium's resonant oscillation through space.

1.3.2   Resonance in Wave Dynamics

The discussion in the preceding section indicates that a standing wave causes the wave energy to stay in the same area in the space. It is easily conceived that by setting up a pair of reflectors we can form a standing wave and thereby confine the wave energy in the space between the reflectors. Such a device is called a resonator and resonators are used in many areas of engineering. An optical resonator (a laser) is an example and will be discussed in detail in a later chapter. Resonators are also used in acoustics. Most musical instruments use resonators [26], e.g., the body of the violin is designed to resonate with the sound generated by the vibration of the strings. The gravitational bar detector known as the Weber bar [27, 28] is designed to detect tiny gravitational wave signals (of the order of 10^{-16} m in displacement) as the vibration of massive aluminum cylinders. The instrument is tuned to resonate at a predicted frequency of gravitational waves.

Designing an actual resonator is not a straightforward task but the basic concept is not complicated. Here we briefly discuss the basic concept using a plane wave. In short, the most important condition is that the wave reflected off one of the reflectors is in phase with the wave reflected from the same reflector after multiple reflections. When a standing wave is formed between the pair of reflectors as schematically illustrated by Fig. 1.11, counter-propagating traveling waves go back and forth in the space between the reflectors. If the phase of a wave that is just reflected off the right reflector is in phase with the one reflected off one round trip before, the two waves interfere with each other constructively. All the waves reflected off the same reflector can be arranged to be in phase with one another.

In the case of a plane wave, the phase is constant over the planar wavefront perpendicular to the axis of propagation. Therefore, the above in-phase condition is straightforward for a pair of planar reflectors. Let l be the distance between the planar reflectors and λ be the wavelength. The phase after each round trip is $2\pi(2l/\lambda)$. If this is equal to an integral multiple of 2π, the in-phase condition is satisfied. Below we consider the condition of resonance for three general cases seen in Fig. 1.11.

1. Fixed-fixed end reflection

   The boundary condition of this case is that at both reflectors the total amplitude is null. Figure 1.11a illustrates the condition where the left-going wave (solid line) is reflected off at the left reflector. The reflected wave (dashed line) makes the total amplitude zero at this reflector. The same boundary condition is established on the right reflector. Assume

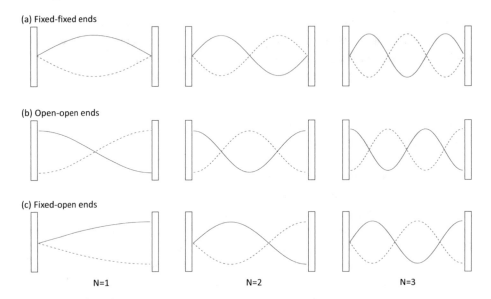

(a) Fixed-fixed ends

(b) Open-open ends

(c) Fixed-open ends

N=1   N=2   N=3

Fig. 1.11 Phase condition for a passive resonator. In the case of an active resonator a source is placed inside the resonator

that the waves are sinusoidal. Since sine and cosine functions take zero at every half periodicity, this condition can be expressed as the resonator length is an integral multiple of half wavelength.

$$l = \lambda \frac{N}{2} \tag{1.108}$$

Here N is an integer. Note that this resonant condition is different from the resonant condition of oscillation where the driving frequency is equal to the natural frequency.

Often our interest is to know the resonant frequency for a given resonator length. Using the general relation $v_p = \lambda f$ (v_p is the phase velocity of and f is the frequency of the wave), we can rewrite the resonance condition (1.108) as follows:

$$f_r^N = \frac{N}{2l} v_p \tag{1.109}$$

Here f_r^N is the Nth resonant frequency.

By letting $z = 0$ and $z = l$ be the coordinates of the left and right reflectors and using the form of Eq. (1.96), we can express the standing wave $\xi_s(t, z)$ as follows:

$$\xi_s(t, z) = 2A \cos\left(\frac{N\pi}{l} z + \frac{\pi}{2}\right) \sin\left(2\pi f_r^N t\right) \tag{1.110}$$

Here the first term on the right-hand side represents the spatial pattern of the standing wave of the Nth order and the second term represents the temporal oscillation at the Nth resonant frequency given by Eq. (1.109).

2. Open-open end reflection

Repeating the same argument except for the boundary conditions, we find the resonant frequency for this case as follows:

$$f_r^N = \frac{N}{2l} v_p \tag{1.111}$$

The corresponding standing wave expression is

$$\xi_s(t, z) = 2A \cos\left(\frac{N\pi}{l} z\right) \sin\left(2\pi f_r^N t\right) \tag{1.112}$$

3. Fixed-open end reflection

In this case, the amplitude is null at the left reflector and at the maximum on the right reflector (Fig. 1.11c). Since the first maximum is a quarter wave away from the first null, and the subsequent maxima are every half wavelength, the resonator length in this condition is equal to a quarter wavelength plus an integer number of half wavelength.

$$l = \lambda\left(\frac{N}{2} - \frac{1}{4}\right) \tag{1.113}$$

Using the phase velocity $v_p = \lambda f$ in (1.113) and solving for the resonant frequency, we find

$$f_r^N = \left(\frac{N}{2l} - \frac{1}{4l}\right) v_p \tag{1.114}$$

When the incoming wave has a resonant frequency, the pair of reflectors act as a resonator. In an ideal resonator, the above-mentioned back-and-forth propagation of the wave continues until the incoming wave is shut off. This means that the wave energy keeps growing inside the resonator. In a real resonator, various loss mechanisms make the waves inside a resonator decay. In other words, the number of round trips for the wave that enters the resonator at a given time is not infinite. If a source provides the resonator with the same amount of energy that the resonator loses with the loss mechanism, a steady state is established and we can store fixed energy in the resonator. Such an energy source can be inside or outside of the resonator. The former is referred to as an active resonator and the latter is as a passive resonator.

1.4   Big Picture of Electromagnetic Wave

In this section, we consider the governing equation of the electromagnetic wave in analogy with the mechanical wave cases. Since an electromagnetic wave is a transverse wave, we can expect that the governing equations of the electromagnetic field are analogous to those of the shear wave dynamics.

As we will discuss in more detail later in this book, Maxwell's equations describe the wave dynamics of the electromagnetic field. Among the four Maxwell's equations, the following two equations are analogous to the shear wave Eqs. (1.102) and (1.103):

$$\nabla \times \mathbf{E} = -\frac{\partial \mathbf{B}}{\partial t} \tag{1.115}$$

$$\nabla \times \mathbf{B} = \epsilon_e \mu_e \frac{\partial \mathbf{E}}{\partial t} \left(+ \mu_e \mathbf{j} \right) \tag{1.116}$$

Here $\mathbf{E}$ and $\mathbf{B}$ are the electric and magnetic field vectors, ϵ_e is the electric permittivity, and μ_e is the magnetic permeability. Equation (1.115) is known as Faraday's law and Eq. (1.116) is known as Ampère's law.

We can replace the displacement vector with the velocity vector in Eqs. (1.102) and (1.103) to clarify the similarity of the shear wave dynamics and electrodynamics.

$$\nabla \times \boldsymbol{\xi} = \boldsymbol{\omega} \rightarrow \nabla \times \mathbf{v} = \frac{\partial \boldsymbol{\omega}}{\partial t} \tag{1.117}$$

$$\nabla \times \boldsymbol{\omega} = -\frac{\rho}{G} \frac{\partial^2 \boldsymbol{\xi}}{\partial t^2} \rightarrow \nabla \times \boldsymbol{\omega} = -\frac{\rho}{G} \frac{\partial \mathbf{v}}{\partial t} \tag{1.118}$$

Comparing Eqs. (1.115) and (1.116) with Eqs. (1.117) and (1.118), we find the following correspondences:

$$\mathbf{v} \rightarrow -\mathbf{E} \tag{1.119}$$

$$\boldsymbol{\omega} \rightarrow \mathbf{B} \tag{1.120}$$

We can discuss the similarity between the mechanical and electromagnetic fields as follows. Equation (1.117) describes the relationship between the longitudinal and transverse degrees of freedom. The electric field is the longitudinal effect and the magnetic field is the rotational effect of the electromagnetic dynamics. The negative sign originates from the restoring mechanism of elasticity. In Eq. (1.118), ρ/G is the reciprocal of the square of the phase velocity. In electrodynamics, Faraday's law describes the restoring mechanism, and $\epsilon_e \mu_e$ is the reciprocal of the square of the speed of light. Table 1.2 summarizes the correspondences.

Table 1.2 Correspondence between shear wave and EM wave

v	ω	G	ρ
E	$-\mathbf{B}$	$1/\mu_e$	ϵ_e

References

1. T. M. Atanackovic, A. Guran, Hooke's Law in *Theory of Elasticity for Scientists and Engineers* 2000th edn. (Birkhäuser Boston, Boston, MA, USA, 2000), pp. 85 - 111
2. G. Mavko, T. Mukerji, J. Dvorkin, *The Rock Physics Handbook* 3rd edn. (Cambridge Univ. Press, New York, 2020), Chap. 2
3. K. J. øAström, R. M. Murray, *Feedback Systems, An Introduction for Scientists and Engineers*, (Princeton Univ. Press, Princeton, London, 2021), pp. 27 -29
4. L. Obando, *Mass-spring-damper system, 73 Exercises Resolved and Explained: Systems Dynamics and Transfer Function*, (Univ. Central de Venezuela, Caracas, Venezuela, 2018)
5. S. Yoshida, *Waves; Fundamental and dynamics* (Morgan & Claypool, San Rafael, CA, USA, IOP Publishing, Bristol, UK, 2017), Chap 1
6. H. Margenau, *The Mathematics of Physics and Chemistry* 2nd edn. (D. Van Nostrand, Co., Inc., Princeton, New Jersey, 1956) Secs. 2.7 and 2.8, pp. 48 - 56
7. G. C. King, *Vibrations and Waves* (Wiley, Chichester, UK, 2009) pp. 33 - 48
8. D. Stipp, *A Most Elegant Equation; Euler's formula and the beauty of mathematics* (Basic Books, New York, USA, 2017)
9. M. L. Boas, *Mathematical Methods in the Physical Sciences* 3rd edn. (Wiley, London, 2006), pp. 61
10. R. N. Bracewell *The Fourier Transform and Its Applications* 3rd edn. (Mc Graw Hill, Boston, New York, 2000)
11. Wolfram MathWorld *Lorentzian Fundction* https://mathworld.wolfram.com/LorentzianFunction.html
12. C. Walck, (Internal Report SUF–PFY/96–01, Univ. Stockholm, 11 December 1996, last modification 10 September 2007) pp. 26 - 35
13. A. J. Sutton Pippard, *Strain Energy Method of Stress Analysis* (Longmans, Green and Co. Ltd., London, 1928)
14. D. B. Weems, *Designing, Building, and Testing Your Own Speaker System with Projecys* 4th edn. (Mc Graw Hill, New York, 1997)
15. "Bulk Modulus", https://eng.libretexts.org/Bookshelves/Civil_Engineering/Book%3A_Fluid_Mechanics_(Bar-Meir)/00%3A_Introduction/1.6%3A_Fluid_Properties/1.6.2%3A_Bulk_Modulus (accessed on August 2, 2022)
16. G. A. Maugin, *Continuum Mechanics through the Ages - From the Renaissance to the Twentieth Century, From Hydraulics to Plasticity* (Springer, New York, 2016)
17. J. N. Reddy, *An Introduction to Continuum Mechanics* 2nd edn. (Cambridge University Press, Cambridge, UK, 2013)
18. J. W. Rundnicki, *Fundamentals of Continuum Mechanics*, (Wiley, Chichester, UK, 2015))
19. L. D. Landau, E. M. Lifshitz, *Theory of Elasticity. Course of Theoretical Physics, vol. 7* 3rd edn. (Butterworth-Heinemann, Oxford, 1986)
20. R. Duga, *History of Mechanics*, Editions du Griffon (Neuchatel, 1955)

21. W. W. Chen, B. Song, *Split Hopkinson (Kolsky) Bar, Design, Testing and Applications* (Springer, New York, 2010)
22. R. H. Randall, *An Introduction to Acousitics* (Dover, Mineola, New York, 2005)
23. G. C. King, *Vibrations and Waves* (Wiley, Chichester, UK, 2009) pp. 137 - 158
24. M. L. Boas, *Mathematical Methods in the Physical Sciences* 3rd edn. (Wiley, London, 2006), pp. 296-298
25. S. Yoshida, *Deformation and Fracture of Solid-State Materials* (Springer, New York, 2015), pp. 94
26. C. S. Jones, *Resonance and Musical Instruments*, https://cnx.org/contents/WnAQOkNR@6/ Resonanceand-Musical-Instruments (accessed on August 2, 2022)
27. J. Weber, Computer Analysis of Gravitational Radiation Detector Coincidences, Nature, **240**, 28-30, 1972
28. O. D. Aguiar, Past, present and future of the Resonant-Mass gravitational wave detectors, Res. Astron. Astrophys **11**, 1, 2011.

Light as EM Wave

2

2.1 Maxwell's Equations

The electric and magnetic fields of a light wave can be conveniently described by the following set of equations known as Maxwell's equations [1–4].

$$\nabla \cdot \mathbf{E} = \frac{\rho}{\epsilon_0} \tag{2.1}$$

$$\nabla \times \mathbf{B} = \epsilon_0 \mu_0 \frac{\partial \mathbf{E}}{\partial t} + \mu_0 \mathbf{j} \tag{2.2}$$

$$\nabla \times \mathbf{E} = -\frac{\partial \mathbf{B}}{\partial t} \tag{2.3}$$

$$\nabla \cdot \mathbf{B} = 0 \tag{2.4}$$

Here $\mathbf{E}$ and $\mathbf{B}$ are the electric and magnetic field vectors, ρ and $\mathbf{j}$ are the density of the electric charge distributed in the medium and its flow (current density), and ϵ_0 and μ_0 are the electric permittivity and magnetic permeability of free space (vacuum). In materials, the electric permittivity is greater than vacuum (see p. 180 of [2]). The magnetic permeability of most non-ferromagnetic materials is close to μ_0 (μ for diamagnetic materials $0 < \mu < \mu_0$, paramagnetic materials $\mu > \mu_0$ and ferromagnetic materials $\mu >> \mu_0$) (see Se. 6.4 of [2]). In this book, ϵ and μ (with no subscript $_0$) are used to mean those constants in materials.

2.1.1 Gauss's Law

Equation (2.1) indicates that the electric charge is the source of the electric field. It is known as Gauss's law. From the definition of divergence, we find that the electric field due to a positive charge is diverging and the field due to a negative charge is converging. Applying the divergence theorem [5], we can put Gauss's law in the integral form.

© The Author(s), under exclusive license to Springer Nature Switzerland AG 2023
S. Yoshida, *Fundamentals of Optical Waves and Lasers*, Synthesis Lectures on Wave
Phenomena in the Physical Sciences. https://doi.org/10.1007/978-3-031-18188-7_2

$$\oint_S \mathbf{E} \cdot \hat{r} \, dS = \iiint_\Omega \nabla \cdot \mathbf{E} \, d\Omega = \frac{1}{\epsilon_0} \iiint_\Omega \rho \, d\Omega = \frac{1}{\epsilon_0} Q \qquad (2.5)$$

Here Q is the total electric charge inside the closed volume Ω. In the case of a point charge, the electric field is spherically symmetric and can be easily evaluated as a function of radial distance from the point charge as follows:

$$\epsilon_0 \mathbf{E} = \frac{Q}{4\pi r^2} \qquad (2.6)$$

Since $4\pi r^2$ is the area of the spherical surface, Eq. (2.6) indicates that the quantity $\epsilon_0 \mathbf{E}$ represents the effect of charge Q averaged over the surface area. This quantity is known as the electric flux density, and literally describes the area density of the diverging (or converging if $Q < 0$) flux due to charge $Q > 0$ at a given distance. As will be discussed later, the electric permittivity in a medium ϵ is higher than ϵ_0. This indicates that in a medium, the electric field strength due to the same charge is weaker than in a vacuum.

We can discuss the concept of electric force based on the electric fluxes exerted by charges. Figure 2.1a illustrates a diverging flux due to a positive point charge, and Fig. 2.1b a converging (negatively diverging) flux due to a negative point charge. The vectors forming the flux represent the electric field lines. Consider that these pairs of positive and negative point charges are placed next to each other. Figure 2.2a illustrates the situation two dimensionally and suggests the nature of their interaction. The diverging nature of the field due to the positive charge and the converging nature of the field due to the negative charge together make the field lines continuous from one charge to the other. We can intuitively imagine that these two unlike charges attract each other as the directions of the fluxes between the charges are facing in the same way. Indeed, unlike charges attract each other. Conversely, like charges repel each other. Figure 2.2b illustrates the situation.

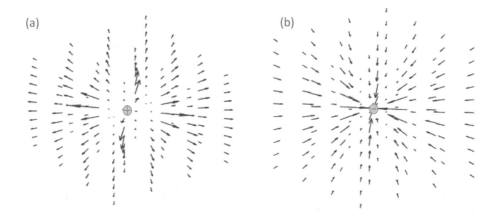

Fig. 2.1 Electric field due to **a** a positive and **b** negative point charge

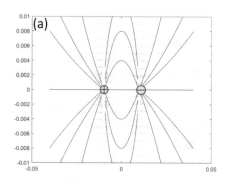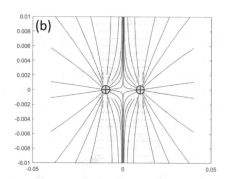

Fig. 2.2 a Electric field due to a pair of positive and negative point charges and **b** Electric field due to a pair of positive point charges

2.1.2   Electric Force

The interaction between electric charges is known as Coulomb force. The magnitude of Coulomb force between a pair of point charges Q_1 and Q_2 is given by the following expression:

$$F_C = k \frac{Q_1 Q_2}{r^2} \tag{2.7}$$

Here k is Coulomb constant, r is the distance between Q_1 and Q_2. When Q_1 and Q_2 have the same polarity ($Q_1 > 0$ and $Q_2 > 0$ or $Q_1 < 0$ and $Q_2 < 0$), $F_C > 0$ represents a repulsive force. When Q_1 and Q_2 are unlike ($Q_1 > 0$ and $Q_2 < 0$ or $Q_1 < 0$ and $Q_2 > 0$), $F_C < 0$ represents an attractive force.

You may notice that Eq. (2.7) is analogous to the universal gravitation equation. If we replace the Coulomb constant with gravitational constant G, and Q_1 and Q_2 with two masses M_1 and M_2, this expression represents the attractive force between the two masses.

$$F_G = G \frac{M_1 M_2}{r^2} \tag{2.8}$$

The only difference is the polarity. Gravitational force is always attractive. Since there is no negative mass, the force expressed by formula (2.8) is always positive, and it is attractive.

The similarity between Coulomb force and gravitational force expressions is not incidental. We can discuss the electric force in a similar fashion to the gravitational force. When we consider the weight of an object of mass m on the surface of the earth, we do not use Eq. (2.8). Instead, we use $f = 9.8 \times m$. This expression is a special case of the following expression:

$$\mathbf{f} = m\mathbf{g} \tag{2.9}$$

Here $\mathbf{g}$ is the gravitational acceleration field vector whose magnitude varies with the distance from the source and the direction toward it. A gravitational field source exerts an acceleration vector field around its center of mass. Comparison of Eqs. (2.8) and (2.9) indicates that if we use the mass of the earth for M_1, the radius of the earth for r and replace M_2 with m, we find $GM_1/r^2 = 9.8$ m/s^2. In this view, mass M_1 is the source for M_2. We can also view M_2 be the source for M_1. In this case, the acceleration field is proportional to M_2.

The use of Eq. (2.9) is convenient to describe the dynamics of an object under the influence of the source's gravitational field. If the object m is under multiple sources of gravity, we can add the gravitational field vectors. For instance, when we describe the dynamics of an object between the earth and the moon, we can simply add the gravitational acceleration vector due to the earth ($\mathbf{g}_e$) and that due to the moon ($\mathbf{g}_m$).

Likewise to the gravitational case, it is sometimes more convenient to use a field vector form like Eq. (2.9) to express the electric force acting on a charge. For instance, this form is more convenient for analyzing the dynamics of an electron orbiting around a proton. The vector expression of the electric force equivalent to Eq. (2.9) takes the following form:

$$\mathbf{f}_e = q\mathbf{E} \tag{2.10}$$

Here $\mathbf{f}_e$ is the electric force exerted by the electric field $\mathbf{E}$ on charge q. Notice that if $q > 0$ the electric force is in the same direction as the electric field, and if the charge is negative the electric force is opposite to the electric field. Expression (2.10) indicates that the unit of the electric field is N/C (Newton per Coulomb).

Similar to the gravitational case, comparison of Eqs. (2.7) and (2.10) tells us the electric field due to a charge Q is

$$\mathbf{E} = k\frac{Q}{r^2}\hat{\mathbf{r}} \tag{2.11}$$

Here $\hat{\mathbf{r}}$ is the unit radial vector. When an electric charge is under the influence of the electric field due to multiple sources, the electric force vector $\mathbf{E}$ in Eq. (2.10) represents the total electric field. In this situation, we can simply add the vector in the form of Eq. (2.11) for each charge (the source charge) exerting an electric field that the charge q (the subject charge) feels.

$$\mathbf{E}_{tot} = \sum_{i=1}^{n} k\frac{Q_i}{r_i^2}\hat{\mathbf{r}}_i \tag{2.12}$$

$$\mathbf{F}_e = q\mathbf{E}_{tot} \tag{2.13}$$

Here Q_i and r_i are the charges of the ith source charge Q_i and its distance from the subject charge q, $\mathbf{E}_i$ is the electric field due to Q_i at the location of q, $\mathbf{E}_{tot}$ is the vector sum of $\mathbf{E}_i$ for the total of n source charges, and $\mathbf{F}_e$ is the total electric force on q due to all the source charges.

2.1.3   Electric Potential Energy and Potential

An electric charge q placed in an electric field $\mathbf{E}$ is under the continuous influence of the electric force. An external agent needs to do some work to displace the charge from a position

r_a to r_b along the field line. We can calculate the work done by the external agent W_{ex} by line-integrating the force it exerts on the charge, $\mathbf{f}_{ex}$.

$$W_{ex} = \int_{r_a}^{r_b} \mathbf{f}_{ex} \cdot d\mathbf{r} \tag{2.14}$$

Here $\mathbf{f}_{ex}$ is the reaction to the electric-field force acting on the electric charge $q\mathbf{E}$ (Eq. (2.10)), i.e., $\mathbf{f}_{ex} = -q\mathbf{E}$. Thus, we can rewrite Eq. (2.14) as follows:

$$W_{ex} = -q \int_{r_a}^{r_b} \mathbf{E} \cdot d\mathbf{r} \tag{2.15}$$

We can interpret W_{ex} as the work done by the external agent to store the potential energy in the electric field. This concept is analogous to that of the gravitational potential energy and the spring potential energy. By lifting a mass to a higher position, we can store the work as gravitational potential energy. Similarly, by compressing a spring we can store our work in the spring's potential energy.

In electrodynamics, it is more common to use electric potential than electric potential energy. The electric potential [6] (also see p. 77–82 of [2]) is defined by the electric potential energy for a unit positive charge.

$$V = -\int_{r_a}^{r_b} \mathbf{E} \cdot d\mathbf{r} \tag{2.16}$$

Here V is the electric potential at point r_b relative to point r_a. Many authors use ΔV or V_{ab} instead of V to write the above expression to clarify the potential is relative to point r_a. Naturally, $V_{ab} = -V_{ba}$. In this book, we use V for simplicity. (The symbol V is used because the electric potential is measured with the unit of volts.) In most engineering cases, the potential difference (the absolute value of V) is important. So, people tend to say "The voltage at point a is higher than point b by 10 V, etc." without using a sign.

When the electric potential is expressed as a function of r, normally we set the potential at infinity to be zero.

$$V(r) = -\int_{\infty}^{r} \mathbf{E} \cdot d\mathbf{r} \tag{2.17}$$

The electric potential can be expressed in the differential form as well.

$$\mathbf{E} = -\nabla V \tag{2.18}$$

Expression (2.18) defines the electric field as the slope of the electric potential at any point in the space. Based on this expression, we can say that the unit of the electric field is V/m (volts per meter).

Unlike the gravitational case, the sign of W_{ex} depends on the sign of q. Since V is defined as the electric potential energy for a positive unit charge, the electric potential for the same initial and final point r_a and r_b acts differently on a negative charge from a positive charge.

For instance, consider that a voltage supply makes 10 V between the ground (0 V) and a metal plate. For a positive charge, the position at the metal plate is higher in the electric potential. Thus, if released, it moves toward the ground. For a negative charge, the ground is higher in potential. In other words, a positive charge will go down the slope of electric potential defined by (2.18) whereas a negative charge climbs up the slope.

2.1.4 Electric Dipole and Interaction with Light

Notice that the situation illustrated by Fig. 2.2 is statically unstable because the two charges keep pulling each other and eventually they are in contact. However, if charges are dynamic, it is possible to have a stable situation. It is similar to the moon's revolution around the earth. If the moon stops revolving, the moon and earth will coalesce into one continuous object. A pair of unlike charges placed close to each other is known as an electric dipole [7]. An electron orbiting an atom (a bound electron) forms an electric dipole with the nucleus. The electric dipole plays an important role in light–matter interaction. This topic will be discussed later in Sect. 3.1.1, but we can intuitively understand the light–dipole interaction using a bound electron as an example.

Depending on the atomic state, the bound electron orbits at a certain frequency. Light is an electromagnetic wave. Consider that a light wave is incident to an orbiting electron. Figure 2.3 illustrates the situation schematically. Imagine that at some moment when the electron is at the top of the orbit, the electric field vector of the light wave is upward, pointing outward away from the nucleus of the atom. Being a negative charge, the electron is pushed by this electric field toward the nucleus. In other words, this electric force acts as an additional centripetal force and accelerates the electron. Assume that the frequency of the light is the same as the orbiting frequency of the electron. At a half period later, when the electron comes to the bottom of the orbit, the electric field vector of the light wave is downward. The electric field exerts a force on the electron toward the nucleus again.

As the light wave keeps exerting the electric force on the electron in this pattern, the oscillatory energy of the electric field is transferred to the kinetic energy of the electron, and hence the electron increases its circular speed. After a certain time, the transferred energy becomes large enough to push the electron to the next outer orbit. This dynamics can be intuitively understood by imagining a small weight connected to an end of a spring undergoing a circular motion. Consider that you hold the other end of the spring and spin yourself so that the weight revolves around you. As you spin faster, the weight increases the circular speed. At the same time, the spring gets stretched because the faster revolution requires a stronger centripetal force. Here the initial length of the spring (the length before you start to spin faster) corresponds to the radius of the initial orbit of the electron and the final length corresponds to the radius of the outer orbit. Note that the spring can increase its length continuously with the spinning speed but the orbit radius cannot change continuously. This is because the atom, as a quantum system, is allowed to possess only discrete electronic

Fig. 2.3 Schematic illustration of interaction between electric field of light wave and orbiting electron

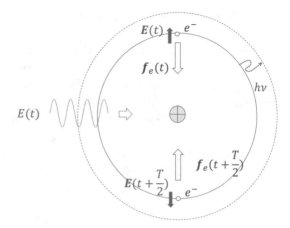

energy levels. In other words, the kinetic energy of the orbiting electron is quantized, and so is the change in the orbit radius. In Fig. 2.3, $h\nu$ represents the quantized energy where h is Planck's constant and ν is the resonant frequency.

This simplified description explains the mechanism known as the photoexcitation [8] of atomic systems. It is clearly a case of resonant interaction between the electric field of light and the dynamics of the atomic system. This explains why atomic emission or absorption spectra exhibits patterns similar to the resonance pattern shown by Fig. 1.11 discussed under "Resonance" in Sect. 1.3.

2.1.5 Ampère's Law with Maxwell's Term

Equation (2.2) is known as Ampère's law with Maxwell's term. As is the case of Gauss's law (2.1), the right-hand side of the equation is the source of the field quantity on the left hand, **B**. Unlike Eq. (2.1), however, Eq. (2.2) indicates that there are two sources of the magnetic field: the temporal derivative of the electric field (the first term) and the electric current (the second term). The first term was added by Maxwell and therefore is called Maxwell's term. The second term represents electric current, the current of electric charges. Notice that this term is a flow of material whereas the first term is the temporal change of the electric field. Sometimes the first term is called the Displacement Current. The phenomenon that an electric current generates a magnetic field is known as Ampère's law. The phenomenon that an electric current or a temporally varying electric field generates a magnetic field is known as Ampère–Maxwell's law. So we can say that Eq. (2.2) represents Ampère–Maxwell's law with the two terms on the right-hand side, and Ampère's law without the first term on the right-hand side.

Magnetic field due to flow electric current

Consider first the electric current as a source of the magnetic field. According to Eq. (2.2), the magnetic field is generated around the source current where the curl of the magnetic field is proportional to the current density. This is comparable to the case of the electric field generated around a point source where the divergence of the electric field is proportional to the charge density. However, note that the current density as the source of the magnetic field is a vector quantity, unlike the electric charge density is a scalar quantity. The quantity $\mathbf{j}$ that appears on the right-hand side of Eq. (2.2) is a vector representing the current density. Its direction is the flow of positive charges. With the charge density, ρ (C/m^3), and the drift velocity of the charges $\mathbf{W}_d$ (m/s) the current density can be expressed as follows:

$$\mathbf{j} = \rho \mathbf{W}_d \qquad (2.19)$$

The unit of $\mathbf{j}$ is (C/m^3)×(m/s)=(C/s)×m/(m^3)=A/m^2. The direction of $\mathbf{j}$ is the same as $\mathbf{W}_d$ if $\rho > 0$ and opposite if $\rho < 0$. For instance, the direction of $\mathbf{j}$ is the same as the flow of positive ions, and opposite to the flow of electrons.

The integral form of Ampère's law, corresponding to the integral form of Gauss's law (2.5), is given as follows:

$$\oint_C \mathbf{B} \cdot d\mathbf{r} = \iint_S (\nabla \times \mathbf{B}) \cdot \hat{\mathbf{n}}\, dS = \mu_0 \iint_S \mathbf{j} \cdot \hat{\mathbf{n}}\, dS = J \qquad (2.20)$$

Here Stokes' theorem [9] is used in going through the first equal sign and Ampère's law is used in going through the second equal sign. $\hat{\mathbf{n}}$ is the unit vector normal to the cross-sectional area. Figure 2.4 illustrates $\mathbf{j}$ vectors due to positive charges flowing in an infinitely long and infinitesimally thin wire forming the total current J.

Total current J generates a magnetic field outside the wire as shown in Fig. 2.4. According to Eq. (2.20), the magnetic field is cylindrically symmetric about the linear source $\mathbf{j}$ and its direction is such that $\nabla \times \mathbf{B}$ is parallel to $j\hat{\mathbf{n}}$ vector. When the wire is infinitely long and its radius is negligibly small, the magnitude of the magnetic field is given as follows:

Fig. 2.4 Schematic illustration of current density $\mathbf{j}$ forming total current J

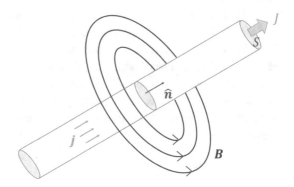

$$B_{out} = \frac{\mu_0 J}{2\pi r} \tag{2.21}$$

Here μ_0 is the magnetic permeability in free space and r is the radial distance from the wire. Although Fig. 2.4 illustrates the concentric pattern at one point along the length of the wire, the magnetic field is uniform along the entire length of the wire.

Magnetic field due to displacement current

Next, consider the first term on the right-side hand of Eq. (2.2), the displacement current, as the source of a magnetic field. The following argument will help us understand that the displacement current generates a magnetic field in a similar fashion to the electric current. (This argument is similar to the one made by Maxwell to modify Ampère's law. [10, 11]. See Appendix .) Equation (2.19) tells us the current density is a flow of charge in a unit volume. Consider in Fig. 2.5 that a swarm of positive charges are injected in the gap of a parallel plate capacitor. The positive charges are injected in a small volume Ω and the total charge is $q_0 = \rho \times \Omega$. At the moment when the positive charges are injected, the circuit switch is turned on so that some voltage is applied across the capacitor's electrodes. The swarm of the positive charges drifts toward the cathode. Around the charges, a magnetic field will be generated as illustrated in Fig. 2.4. What is the situation in other parts of the gap? In the circuit wire, charges flow continuously due to the voltage difference. The same amount of charges enter the wire through the cathode and exit the wire through the anode. This means that in a space near the cathode there must be something like a current flowing. This is the displacement current.

We can understand the meaning of the displacement current term as follows. According to Eq. (2.1), the movement of positive charges generates a diverging electric field of strength inversely proportional to the distance from the center of the swarm. In Fig. 2.5, we can easily see that in front of the swarm, the electric field increases as the swarm approaches because it carries the diverging electric field. As the distance between the point of interest and the center of the swarm diminishes, the electric field becomes greater at that point. Conversely, the electric field decreases behind the swarm as it drifts away. These temporal changes are accounted for by $\partial \mathbf{E}/\partial t$, which appears on the right-hand side of Eq. (2.2). Clearly, this term

Fig. 2.5 Swarm of positive charges drift in gap of capacitor generating magnetic field

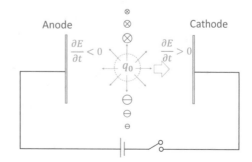

is comparable to the current density term $\mu_0\mathbf{j}$. Here, we need to multiply the term by ϵ_0 to convert the dimension from the electric field to the charge (see Eq. (2.1)) and by μ_0 as we do for the $\mathbf{j}$ term.

Notice that the magnetic field is cylindrically symmetric about a thin current source whereas the electric field is spherically symmetric about a point charge. These indicate that the electric field is divergence-like and the magnetic field is rotation (curl)-like vector fields. This difference leads to the fact that the magnetic force does not do work on charges while the electric field does work on charges, as will be discussed in the next section.

2.1.6   Magnetic Force

In the above section, we introduced electric force as the interaction between electric charges. Since an electric charge is the source of an electric field, we can interpret the electric force as the interaction between the field and its source. The field expression (2.10) explicitly indicates the interaction. The magnetic force also can be expressed as the product of a field and source. However, there is some difference from the electric counterpart. In the electric case, the charge is a scalar. Hence, the electric force as the product of the electric field and charge is always in line with the electric field. The direction of the force is solely determined by the sign of the charge. As we discussed above, the source of a magnetic field is a current density, which is a vector quantity. Since the magnetic field is a vector and the force is also a vector, this product must be a vector product. The magnetic force expression corresponding to the electric force expression (2.10) can be given by one of the following:

$$\mathbf{f}_{\mathbf{Bq}} = q\mathbf{v} \times \mathbf{B} \tag{2.22}$$

$$\mathbf{f}_{\mathbf{Bj}} = J\mathbf{l} \times \mathbf{B} \tag{2.23}$$

Expression (2.22) represents the magnetic force acting on a charge moving with velocity $\mathbf{v}$. Expression (2.23) represents the magnetic force acting on length l part of a wire carrying electric current J. Vector $\mathbf{l}$ indicates the direction of current J.

Figure 2.6 illustrates a top view of two wires carrying electric current in the out-of-page direction. Both wires act as a source of a magnetic field forming counterclockwise concentric vectors. Note that at the location of the right wire, the magnetic field due to the left current is upward. Since the current flows in the out-of-page direction, the direction of vector $\mathbf{l}$ is out-of-page. Hence, the vector product $\mathbf{l}$ and $\mathbf{B}$ is to the left, i.e., toward the other current. We can repeat the same argument for the left wire. There the magnetic field due to the right current is downward whereas the current carried by the left wire is out-of-page. Therefore, the vector product indicates that the magnetic force acting on the left wire is rightward, i.e., toward the right wire. This analysis reveals that a pair of wires carrying electric current in the same direction attract each other.

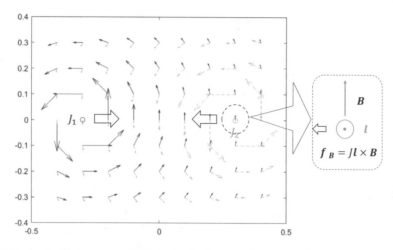

Fig. 2.6 Magnetic fields generated by a pair of wires carrying current in the same direction. The magnetic force is attractive. The illustration on the right shows the magnetic field and force vectors in the area highlighted by the dashed-circle

As you will easily imagine, if the directions of the currents are opposite, the two wires repel each other. Figure 2.7 illustrates the situation.

The direction of the magnetic force resulting from the vector product has another significance. Consider the direction of the magnetic force with the second expression (2.23). It is perpendicular to the velocity, hence the magnetic field does not do any work. It is similar

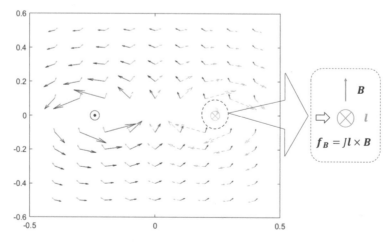

Fig. 2.7 Magnetic fields generated by a pair of wires carrying current in opposite directions. The magnetic force is repulsive. The illustration on the right shows the magnetic field and force vectors in the area highlighted by the dashed-circle

to the situation where centripetal force does not do work when it causes a circular motion at a constant speed. On the contrary, being parallel to the electric force, the electric field does do work. In the context of electromagnetic wave dynamics, usually, the work done by an electric field via the electric force acting on a free charge dissipates the electromagnetic energy and thereby causes the electromagnetic wave to decay.

2.1.7   Magnetic Dipole

Considering the existence of electric dipoles, it is natural to speculate if magnetic dipoles exist. The answer is yes. Physically a magnetic dipole [12] can be produced by a pair of electric currents flowing in opposite directions like the one illustrated in Fig. 2.7. Look at the space between the two currents. There the magnetic fields produced by the two currents are pointing in the same direction vertically, and in the opposite directions horizontally. This pattern is somewhat similar to the electric dipole illustrated in Fig. 2.2 where the field vector is directed from the positive to the negative charge of the dipole. In the case of Fig. 2.7, the vectors are pointing upward, so the situation corresponds to the pattern you would see by rotating Fig. 2.2 by 90° counterclockwise. Figure 2.8 illustrates the total vector field due to the two currents.

However, there is a crucial difference between the electric and magnetic dipoles. Notice that in the case of the electric dipole, the field lines originate from the positive charge and end at the negative charge. In the case of the magnetic dipole, there are no starting points or ending points of the field lines as they curl around the sources. In other words, there is no diverging or converging pattern in the magnetic field. This very fact is beautifully

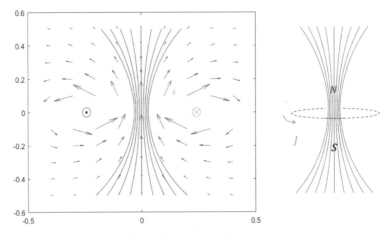

Fig. 2.8 Magnetic dipole. "N" and "S" denote North and South poles of the magnet

Fig. 2.9 An electric magnet and a bar magnet

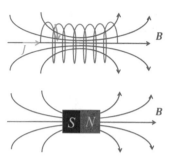

represented by the fourth Maxwell's Equation (2.4), "the magnetic field is divergenceless". Sometimes this fact is expressed by the phrase "there are no magnetic monopoles" [13].

The right picture in Fig. 2.8 illustrates that a loop current can produce a magnetic dipole. Here the current flows out of the page on the left side of the loop and flows into the page on the right side. We can view this as one loop of wire in a solenoid, which normally consists of hundreds of loops stacked on one another as schematically illustrated in Fig. 2.9. The solenoid functions as an electromagnet when a current flows through. The pole where the field line leaves the dipole is called the North (N) pole, and the pole where the field line enters the dipole is called the South (S) pole. Figure 2.9 shows a permanent (bar) magnet that has the same pole arrangement as the electromagnet.

2.1.8 Faraday's Law and Maxwell's Term as Lenz's Law

Equation (2.3) is known as Faraday's law [14] and is extremely important in the wave dynamics of the electromagnetic field. In Sect. 1.1, we discussed that a wave is the spatial transmission of a temporal oscillation. Faraday's law translates the temporal change (the right-hand side) to the spatial change (the left-hand side). From this standpoint, we can view Faraday's law as constituting the mechanism that generates the wave dynamics in the electromagnetic field. However, Faraday's law alone does not generate the restoring mechanism essential for the wave dynamics. The restoring mechanism is produced by the synergetic interaction between the electric and magnetic fields resulting from a combination of Faraday's law and Maxwell's term in Ampère–Maxwell's law. In the next few paragraphs, we discuss the synergetic interaction that produces the restoring mechanism of the electromagnetic field.

Restoring mechanism initiated by temporal change of magnetic field

Figure 2.10 illustrates the mechanism in a scenario where a temporal change in the magnetic field induces a spatial change in the electric field. This phenomenon is known as electromagnetic induction. Consider that a bar magnet moves leftward toward the observation point

Fig. 2.10 Restoring
mechanism initiated by
temporal change in magnetic
field

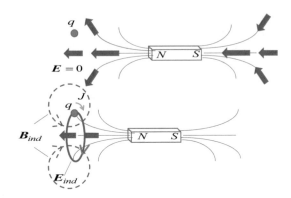

where an observer holds a positive charge q. Initially, the magnet is so far away that the
magnetic field at the observation point is negligibly small. As the magnet approaches, the
magnitude of the magnetic field becomes significant at the observation point, and the charge
q starts to feel the temporal increase in the magnetic field. According to Faraday's law,
this temporal change in the magnetic field induces a circular electric field as depicted by
the solid circle in Fig. 2.10. Consequently, the charge q moves along the circular path. Pay
attention to the direction of the induced electric field. It is counterclockwise viewed from
the magnet's side. This direction is determined by the sign of $\partial \mathbf{B}/\partial t$ and the negative sign
on the right-hand side of Eq. (2.3).

This induction of the electric field produces a displacement current, $\epsilon_0 \mu_0 \partial \mathbf{E}/\partial t$, flowing
along the circular path as well. According to Ampère–Maxwell's law, this displacement
current produces a circulating magnetic field as depicted by the dashed circular lines in Fig.
2.10. This time there is no negative sign on the right-hand side. Hence, the direction of the
produced magnetic field is as indicated by the dashed circular lines. Notice that this direction
is opposite to the increasing magnetic field due to the bar magnet and its movement. In other
words, the magnetic field produced by the induced electric field (the secondary magnetic
field) opposes the change in the magnetic field due to the bar magnet (the primary magnetic
field). This opposing effect is generated by the negative sign on the right-hand side of Eq.
(2.3). Without the negative sign, the direction of the induced electric field would be opposite,
hence the direction of the secondary magnetic field would enhance the initial change in the
primary magnetic field. It is clear that such a system is unstable; once the primary magnetic
field increases over time due to the motion of the magnet, it keeps increasing.

If the primary magnetic field decrease over time at the observation point, the temporal
change in the magnetic field becomes negative on the right-hand side of Eq. (2.3) ($\partial \mathbf{B}/\partial t <$
0), hence the left-hand side of the equation becomes positive. This causes the charge q to
move and the displacement current to flow opposite to the situation depicted in Fig. 2.10,
and thereby the secondary magnetic field becomes leftward. Again, the secondary magnetic
field opposes the change in the primary magnetic field.

This type of opposing effect is generally known as Lenz's law [15]. We can argue the opposing effect in force as well. In the first situation discussed above, the primary magnetic field increases leftward and the secondary magnetic field is rightward. Referring to Fig. 2.9 we find that having the rightward secondary magnetic field is equivalent to placing a small bar magnet with the N pole on the right and the S pole on the left. Note that this N pole of the secondary bar magnet faces against the N pole of the primary bar magnet. This N-vs-N pole situation repels the two bar magnets away from each other. The secondary bar magnet literally opposes the primary bar magnet's movement.

We can repeat the same argument for the case when the primary magnet moves away from the observation point. In this case, the secondary bar magnet has the S pole on the right attracting the N pole of the primary magnet. Again, the magnetic force opposes the motion of the primary magnet.

In the above event, the motion of the charge q involves the work done by the induced electric field, causing dissipation in the electromagnetic field. However, the displacement current does not cause energy dissipation. In other words, this restoring mechanism can continue forever if there is no electric charge in the system. This situation corresponds to the generation of non-decaying electromagnetic waves. In reality, an electromagnetic wave always decays. However, this observation explains why our cell phone's signal is weaker in a building. The free charges existing in the building material, such as the steel frame, cause the electromagnetic energy to dissipate.

Restoring mechanism initiated by temporal change of electric field

In the above scenario, the synergetic interaction was initiated by the temporal change in the magnetic field. In the next scenario, we consider the same synergetic interaction using an electric dipole. In this scenario, the interaction is initiated by a motion of a small number of positive charges. The motion induces a temporal change in the electric field, which triggers the synergetic interaction and moves the nearby negative charges. Consider the illustration in Fig. 2.11. The top disk containing positive charges represents the positive charge of a dipole and the other disk represents the negative charge. Initially, the two disks are held by an external agent. When the agent releases the disks, the Coulomb force pulls the two disks toward each other. This increases the electric field in the gap between the disks, and consequently, a magnetic field is produced in accordance with Ampère–Maxwell's law. The solid loop in Fig. 2.11b depicts the produced magnetic field. Since this magnetic field is newly produced in the gap, its temporal change is positive. Then, according to Faraday's law (2.3), an electric field is induced as depicted by the dashed line in Fig. 2.11b. Apparently, this electric field exerts an electric force that separates the two disks.

The above illustrated opposing effect is the source of a restoring force. When we solve Maxwell equations by combining Faraday's law and Ampère–Maxwell laws in the next section, we can algebraically derive a wave equation. Before deriving an electromagnetic wave equation, it is worthwhile to discuss the contrastive natures of the electric and magnetic

Fig. 2.11 Restoring mechanism initiated by temporal change in electric field

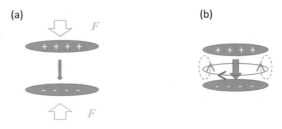

fields through analysis of Faraday's law and Maxwell's term. Dimensional analysis of Faraday's law (2.3) indicates that $[E]/m=[B]/s$, i.e., $[E]/[B]$ has the dimension of velocity m/s. Here $[E]$ and $[B]$ denote the dimensions of the electric and magnetic fields, respectively. Dimensional analysis of the left-hand side and Maxwell term of Ampère–Maxwell laws (2.2) indicate that $[B]/m=[(\epsilon_0\mu_0)E]/s$, i.e., $[E]/[B]=(s/m)/[(\epsilon_0\mu_0)]$, which according to the dimensional analysis on Faraday's law must be in m/s. It follows that $[(\epsilon_0\mu_0)]=(s/m)^2$. In other words, the quantity $1/(\epsilon_0\mu_0)$ must have the dimension of velocity squared. In fact, $1/\sqrt{\epsilon_0\mu_0}$ is the speed of light in free space. This topic will be discussed again later in this chapter (Sect. 2.2).

2.2   Wave Equation and Solutions

2.2.1   Wave Dynamics of Electromagnetic Fields

The wave dynamics of the electromagnetic field can be understood through a consideration of the spatiotemporal behavior of the magnetic and electric fields in the surrounding area based on Ampère's law (2.2) and Faraday's law (2.3). Suppose that an external agent causes an initial change in the magnetic field at a point in space. Consider several time steps after this initial, temporal change in Fig. 2.12. In this figure, the time evolves downward from the top row with an increment of Δt. Columns labeled ΔB and ΔE represent the change in the respective fields at each time segment, and column labeled B represents the magnetic field at the spatial point where the initial agent causes the initial change. In each time step, dashed field lines indicate the cause of the change and the solid field lines indicate the resultant change. For instance, the illustration in row $t + \Delta t$ and column ΔE represents the event where the initial ΔB caused by the external agent induces the electric field. Similarly, the illustration in row $t + 2\Delta t$ and column ΔB represents that the electric field induced at time step $t + \Delta t$ induces magnetic field in time step $t + 2\Delta t$. The two right columns indicate the situation that at the point where the initial change in the magnetic field occurs, the magnetic and electric fields alternate their direction at every other time step. This infers that the function describing the magnetic and electric fields alternates their signs through second-

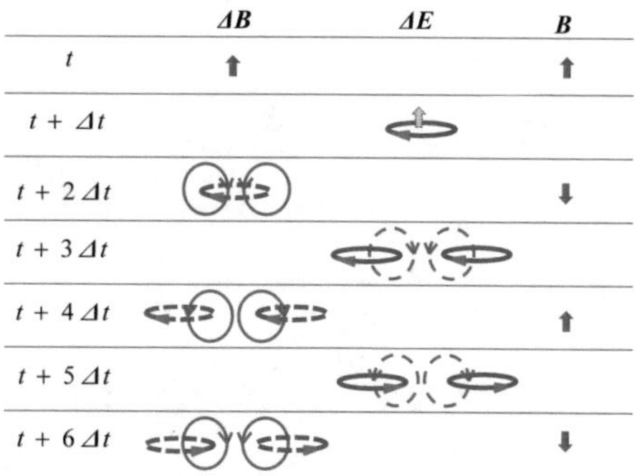

Fig. 2.12 Time evolution of induced magnetic and electric fields

order temporal differentiation keeping the type of function the same, i.e., $\partial f^2/dt^2 \propto f$ where f denotes the function. We know that sine and cosine functions have this property. The alternation of the sign (the direction) of each field corresponds to the fact that sine and cosine functions change their signs during every phase change of π.

The two columns labeled ΔB and ΔE also indicate that the temporal change in the magnetic and electric fields induces the rotational change of the other field in the next time step. This reflects the mathematical relation we observed in Eqs. (2.3) and (2.2); $/\partial t$ of one field variable is proportional to $\nabla \times$ of the other field. This synergetic interaction between the two fields generates the oscillatory behavior of the respective fields that propagates as a wave.

2.2.2 Pictorial Explanation of Electromagnetic Waves

Refer to Fig. 2.13 and consider the spatiotemporal oscillatory characteristics of the electromagnetic wave dynamics in more detail. This figure illustrates the transverse spatial behavior of ΔB and ΔE shown in columns in Fig. 2.12. Here, Fig. 2.13a illustrates the situation where the effect of electromagnetic induction spreads out in a transverse direction at each time step. The inducing and induced fields are depicted in the same format as shown in Fig. 2.12. Figure 2.13b illustrates the direction of the induced field at each time step.

Consider a field induced at a time step, for instance, the electric field at time step $t + 3\Delta t$. The temporal change of this "induced" field "induces" a magnetic field at the next time step $t + 4\Delta t$. Notice that the induction of the magnetic field at this time step occurs over a wider transverse range than at time step $t + 2\Delta t$. This is because the magnetic field is induced continuously over space including the outer sides of the outermost pair of the ring-shaped electric field lines. In this fashion, with the passage of time, the induced electric and magnetic fields spread out from the initial point where the external agent triggers the effect.

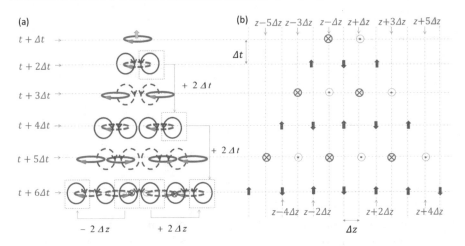

Fig. 2.13 Time and space evolution of induced magnetic and electric fields

Now pay attention to the pattern of the outermost induced magnetic field, the magnetic field enclosed by squares in Fig. 2.13a. We can find that moving forward in time by $2\Delta t$ and moving in space by $\pm 2\Delta z$ bring back the same pattern of the magnetic induction. Figure 2.13b illustrates the pattern more explicitly. Moving vertically along the time axis and horizontally along the space axis, we find the same alternating behavior in both the electric and magnetic fields. Mathematically, this means that the temporal secondary differentiation of the wave function is proportional to the spatial secondary differentiation of the wave function, i.e., the wave equation of the electromagnetic field has the following form:

$$\frac{\partial^2 f}{\partial t^2} = c^2 \frac{\partial^2 f}{\partial z^2} \tag{2.24}$$

As will be discussed in more detail, the constant of proportionality c^2 in the differential equation (2.24) is the square of the phase velocity of the electromagnetic wave

$$c = \frac{1}{\sqrt{\epsilon \mu}} \tag{2.25}$$

known as the speed of light. The physical meaning of $1/\sqrt{\epsilon \mu}$ as the interaction time of the electromagnetic wave with the medium that the light is passing through will be discussed in the next section.

While the phase velocity of electromagnetic waves is discussed in detail in the following section, it is instructive to discuss it intuitively as follows. Figures 2.12 and 2.13 indicate that the temporal change in the electric field induces a magnetic field through the spatial differentiation $\nabla \times \mathbf{B}$. Equation (2.2) has the product $\epsilon \mu$ on the right-hand side multiplied to the temporal derivative of the electric field. This means that the greater the value of $\epsilon \mu$, the smaller the spatial derivative $\nabla \times \mathbf{B}$ for the same temporal change in the electric field. Naively speaking, the spatial derivative of a wave function represents the spatial periodicity or the wavelength. So, this argument indicates that the greater the $\epsilon\mu$, the shorter the wavelength for the same temporal differentiation of the wave function, which corresponds to the temporal periodicity or the frequency. The phase velocity of a wave is the product of wavelength and frequency. So the argument made here can be understood as follows. The greater the value of $\epsilon\mu$, the shorter the wavelength for the same frequency, and the smaller the phase velocity.

2.2.3 LC Oscillation as a Spring-like Mechanism

The physical meaning of the phase velocity $c = 1/\sqrt{\epsilon\mu}$ as being related to the interaction time of the electromagnetic field with the medium that the electromagnetic wave is passing through can be understood through a consideration of an LC circuit and the electromagnetic wave generated by the LC oscillation [16]. Here L stands for the inductance of an inductor (solenoid) and C for the capacitance of a capacitor. The inductance is defined as the constant of proportionality between the voltage across an inductor and the temporal derivative of the current flowing through the inductor. Capacitance is defined as the constant of proportionality between the voltage across a capacitor and the charge stored by the capacitor. More detailed descriptions are found below under "Magnetic energy density" and "Electric energy density", respectively.

Consider Fig. 2.14 that illustrates an ideal (lossless) LC circuit oscillating at the resonant frequency. When the current is in the increasing phase downward inside the solenoid, the inductive reactance generates a magnetic field as shown in Fig. 2.14. This magnetic field changes over time, and the temporal change induces an electric field in area A as shown in the figure. The temporal change in this electric field in turn induces a magnetic field as shown in Fig. 2.14 with a dashed line in area B. In this fashion, magnetic and electric fields are induced one after another in space. This phenomenon is known as the LC oscillation and can be described by the following differential equation:

Fig. 2.14 Electromagnetic fields in an LC circuit

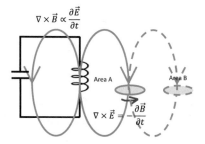

$$L\frac{di(t)}{dt} = -\frac{q(t)}{C} \tag{2.26}$$

where L is the inductance, C is the capacitance, $i(t)$ is the current flowing into the capacitor, and $q(t)$ is the charge stored in the capacitor. Knowing that $i(t) = dq(t)/dt$, we can rewrite Eq. (2.26) into the following differential equation:

$$\frac{d^2q(t)}{dt^2} = -\frac{q(t)}{LC} \tag{2.27}$$

We can easily solve differential equation (2.27) to find an oscillatory solution with angular frequency $1/\sqrt{LC}$. We know that the following solution satisfies differential equation (2.27):

$$q(t) = q_0 \sin \omega t \tag{2.28}$$

Substitution of solution (2.28) into differential equation (2.27) yields

$$-\omega^2 q_0 \sin \omega t = -\frac{1}{LC} q_0 \sin \omega t \tag{2.29}$$

So,

$$\omega = \frac{1}{\sqrt{LC}} \tag{2.30}$$

This frequency is known as the resonant frequency of LC oscillation. We can intuitively understand the reciprocal dependence of the resonant frequency on L and C as follows. The greater the capacitance C, the longer it takes to charge up the capacitor. The greater the inductance L, the more resistive to change the polarity of the current. Hence, the oscillation becomes slower.

We can find the voltage across the capacitor $v_c(t)$ and inductor $v_L(t)$ as follows:

$$v_C(t) = \frac{1}{C}q(t) = \frac{1}{C}q_0 \sin \omega t \tag{2.31}$$

$$v_L(t) = L\frac{di}{dt} = L\frac{d^2q}{dt^2} = -L\omega^2 q_0 \sin \omega t = -\frac{1}{C}q_0 \sin \omega t \tag{2.32}$$

Here Eq. (2.30) is used in Eq. (2.32). The energies stored in the capacitor and inductor are

$$W_{cap} = \frac{1}{2}Cv_C^2 = \frac{q_0^2}{2C} \sin^2 \omega t \tag{2.33}$$

$$W_{ind} = \frac{1}{2}Li^2 = \frac{1}{2}L\left(\frac{dq}{dt}\right)^2 = \frac{L\omega^2}{2}\cos^2 \omega t = \frac{q_0^2}{2C}\cos^2 \omega t \tag{2.34}$$

Equations (2.33) and (2.34) indicate that the capacitive and inductive energies alternate. When one is zero the other is maximum, and vice versa. They also indicate that the maximum capacitive energy is the same as the maximum inductive energy.

Fig. 2.15 Electromagnetic wave propagating in space interacting with effective solenoids and capacitors

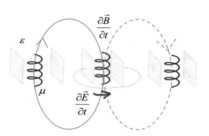

With this LC circuit picture, we can qualitatively explain the mechanism by which electromagnetic waves propagate in space at the velocity $1/\sqrt{\epsilon\mu}$. Free space has capacitance and inductance and can be modeled as a series of capacitors and solenoids (call the effective capacitor and solenoid) [17]. Consider effective capacitors and solenoids in the space in which the electromagnetic wave propagates in Fig. 2.15. When the electric field changes over time, a magnetic field is induced. This is equivalent to the situation where a current flows through the effective solenoid at the point of the space. When the current changes over time, the solenoid induces resistive voltage to suppress the current change in proportion to the inductance. The induced voltage, on the other hand, stores charges in the space to the effective capacitor. Thus, with the effective solenoid and capacitor, the electromagnetic field causes LC oscillation in space.

In this mechanism, L is proportional to μ and C is to ϵ (see next section). Therefore, the greater the product $\epsilon\mu$, the lower the frequency. Consequently, it takes longer for the alternating electromagnetic field to interact with the LC pair at each point of the space, and the electromagnetic wave travels more slowly. Here the storage time is related to charge Q via ϵ and the inductive resisting voltage is related to the first-order temporal derivative of the current, i.e., the second-order temporal derivative of the charge. Thus, the interaction time is determined by the second-order temporal differentiation and the combination of ϵ and μ. Naively speaking, this is why the electromagnetic wave is a solution to the second-order differential equation of **E** or **B** and its phase velocity is related to the product of ϵ and μ. Later in the Sect. 2.2.8 we will discuss that as an electromagnetic wave propagates, electric and magnetic energies are stored in the space alternately.

2.2.4 Electric Energy Density

Consider a parallel plate capacitor [18] and find the electric force acting on the electrode by analyzing how much work is necessary to separate the anode from the cathode for dl against the electric force exerted by the cathode. This method to find a field force is based on the principle of virtual work [19]. Apply Gauss's law in the integral form (2.5) to the cathode that holds a total charge of $-Q$.

$$-Q = \epsilon_0 \iint \mathbf{E} \cdot d\mathbf{a} = 2\epsilon_0 A E \tag{2.35}$$

Here A is the area of the cathode and the thickness of the cathode is omitted from the surface integral.[1] So, the electric field vector is

$$\mathbf{E}_c = \frac{Q}{2\epsilon_0 A}(-\hat{\mathbf{j}}) \tag{2.36}$$

Here $\hat{j}$ is the unit vector whose positive direction is toward the anode.

From Eq. (2.36), we find the electric force exerted by the cathode on $+Q$ held by the anode is

$$\mathbf{F}_c = Q\mathbf{E}_c = \frac{Q^2}{2\epsilon_0 A}(-\hat{\mathbf{j}}) \tag{2.37}$$

The work necessary to pull the anode against $\mathbf{F}_c$ is

$$dW_e = -\mathbf{F}_c \cdot \hat{\mathbf{j}} dl = \frac{Q^2}{2\epsilon_0 A}(\hat{\mathbf{j}}) \cdot \hat{\mathbf{j}} dl = \frac{Q^2}{2A\epsilon_0} dl \tag{2.38}$$

Here $-\mathbf{F}_c$ is the reaction to the electric force exerted by the cathode.

From Eq. (2.38), we can find the total electric energy W_e stored in a parallel plate capacitor with a gap length d as follows:

$$W_e = \int dW_e = \int_0^d \frac{Q^2}{2A\epsilon_0} dl = \frac{Q^2}{2A\epsilon_0} \int_0^d dl = \frac{Q^2 d}{2A\epsilon_0} \tag{2.39}$$

Above, we consider the work done by an external agent to form a parallel plate capacitor by separating the anode from the cathode by the gap distance d. We next change the perspective and consider what the capacitor formed by the external agent can do to a positive test charge placed in a gap just inside the anode. For this analysis, we need to find the total electric field in the gap. Repeating the same analysis as Eqs. (2.35) and (2.36), apply Gauss's law to the anode to obtain the electric field due to Q stored in the anode.

$$\mathbf{E}_a = \frac{Q}{2\epsilon_0 A}(-\hat{\mathbf{j}}) \tag{2.40}$$

Note that the electric field due to the anode is identical to that due to the cathode. Thus, the total electric field in the gap becomes as follows:

$$\mathbf{E}_g = \mathbf{E}_a + \mathbf{E}_c = \frac{Q}{\epsilon_0 A}(-\hat{\mathbf{j}}) \tag{2.41}$$

---

[1] Note that we are considering the electric force on the anode exerted by the cathode. Therefore, we need to use the electric field produced by the charge on the cathode. This electric field is different from the electric field of the capacitor; once the cathode–anode pair forms a capacitor the electric field outside the capacitor becomes null as the positive and negative charges cancel each other.

The electric potential at the anode is

$$V_g = -\int_0^d \frac{Q}{\epsilon_0 A}(-\hat{\mathbf{j}}) \cdot \hat{\mathbf{j}} dl = \frac{Qd}{\epsilon_0 A} \tag{2.42}$$

This is the potential for the test positive charge placed just inside the anode. At this position, the potential is the highest for the test change.

Notice that this maximum potential is always proportional to the charge Q stored in the capacitor. We can express this proportionality using a constant of proportionality C known as the capacitance.

$$Q = \frac{\epsilon_0 A}{d} V_g = CV_g \tag{2.43}$$

$$C = \frac{\epsilon_0 A}{d} \tag{2.44}$$

Note that the capacitance C is proportional to the electric permittivity, and therefore a material-dependent constant.

Substituting Eqs. (2.43) and (2.44) into Eq. (2.39), we can express W_e as follows:

$$W_e = \frac{Q^2}{2C} = \frac{CV_g^2}{2} = \frac{\epsilon_0 A}{2d} V_g^2 \tag{2.45}$$

From Eqs. (2.41) and (2.42),

$$E_g = \frac{V_g}{d} \tag{2.46}$$

Substitution of Eq. (2.46) into Eq. (2.45) yields

$$W_e = \frac{\epsilon_0 A}{2d} (E_g d)^2 = \frac{1}{2} \epsilon_0 (Ad) E_g^2 \tag{2.47}$$

Since Ad is the volume of the capacitor, the electric energy density becomes

$$w_e = \frac{W_e}{(Ad)} = \frac{1}{2} \epsilon_0 E^2 \tag{2.48}$$

Here we drop the suffix g in Eq. (2.48). Note that we assume the electric field is uniform across the gap.

Above, we started the discussion from Eq. (2.5), which describes the relationship between the charge Q in free space and the electric field. That is why we used the electric permittivity of free space ϵ_0 in the above analysis. When we use a dielectric medium in the gap of a capacitor, we replace ϵ_0 with the material's electric permittivity ϵ.

2.2.5 Magnetic Energy Density

By definition, the magnetic flux is the surface integral of the magnetic field, which can be expressed by the vector potential $\mathbf{A}$.

$$\phi = \iint_S \mathbf{B} \cdot d\mathbf{a} = \iint_S (\nabla \times \mathbf{A}) \cdot d\mathbf{a} = \oint \mathbf{A} \cdot d\mathbf{l} \qquad (2.49)$$

Here vector potential is related to the magnetic field as follows [20] (Also see p. 234–246 of [2]).

$$\nabla \times \mathbf{A} = \mathbf{B} \qquad (2.50)$$

From Faraday's law,

$$\frac{d\phi}{dt} = \iint_S \frac{d\mathbf{B}}{dt} \cdot d\mathbf{a} = - \iint (\nabla \times \mathbf{E}) \cdot d\mathbf{a} = - \oint \mathbf{E} \cdot d\mathbf{l} = -V_L \qquad (2.51)$$

Here Stokes' theorem is used in going through the last equal sign and V_L is the voltage across the solenoid. The negative sign in front of V_L represents the opposing effect of Faraday's law.

Since $B \propto \mu_0 i$ (Ampère's law), $\phi \propto \mu_0 i$, and hence from Eq. (2.51), $V_L \propto \mu_0 di/dt$. The constant of proportionality is inductance L.

$$V_L = L \frac{di}{dt} \qquad (2.52)$$

Note that the inductance L is proportional to the magnetic permeability, and therefore a material-dependent constant.

The current flowing into the solenoid flows against the inductive reactance (the resistance associated with inductance L, see Eq. (2.52)). We can express power as the product of current and voltage difference. So, in this case, the power due to the inductive reactance is

$$P = i V_L = i L \frac{di}{dt} = L i \frac{di}{dt} \qquad (2.53)$$

Since energy is the time integral of power, the maximum energy stored in the solenoid is given as follows. Note that this energy storage in the inductor occurs in the first half cycle of one cycle. In the next half cycle, the stored energy is converted to capacitive energy stored in the capacitor.

$$W_{max} = L \int_{t_{min}}^{t_{max}} \left(i \frac{di}{dt} \right) dt = L \int_{t_{min}}^{t_{max}} \left(\frac{1}{2} \frac{d}{dt} i^2 \right) dt = \frac{1}{2} L \left[i^2 \right]_0^{I_{max}} = \frac{1}{2} L I_{max}^2 \quad (2.54)$$

Here t_{min} and t_{max} are the times when the current flowing into the solenoid is the minimum $I_{min} = 0$ and maximum I_{max}.

Integrating Eq. (2.51)

$$\phi = -\int V_L dt = L \int \frac{di}{dt} = L I_{max} \tag{2.55}$$

Replacing $L I_{max}$ with ϕ and using Eq. (2.49), we can express W_{max} (Eq. (2.54)) with vector potential.

$$W_{max} = \frac{1}{2} I_{max}(L I_{max}) = \frac{1}{2} I_{max}\phi = \frac{1}{2} \oint I_{max}\mathbf{A} \cdot d\mathbf{l} = \frac{1}{2} \oint \mathbf{A} \cdot \mathbf{I_{max}} dl$$

$$= \frac{1}{2} \oint \mathbf{A} \cdot \mathbf{j_{max}} dS dl = \frac{1}{2} \iiint_{\Omega} \mathbf{A} \cdot \mathbf{j_{max}} d\Omega \tag{2.56}$$

Note that the current flows along the solenoid coil and therefore in line with dl. $\mathbf{j}$ is the current density and dS is the cross-sectional area of the wire, $d\Omega = dS dl$.

Using Ampère's law and Stokes' theorem,

$$W_{max} = \frac{1}{2} \iiint_{\Omega} \mathbf{A} \cdot \mathbf{j_{max}} d\Omega = \frac{1}{2} \iiint_{\Omega} \frac{1}{\mu_0} \mathbf{A} \cdot (\nabla \times \mathbf{B}) \, d\Omega \tag{2.57}$$

Using the following identity, we can rewrite Eq. (2.57).

$$\nabla \cdot (\mathbf{A} \times \mathbf{B}) = \mathbf{B} \cdot (\nabla \times \mathbf{A}) - \mathbf{A} \cdot (\nabla \times \mathbf{B})$$

$$W_{max} = \frac{1}{2\mu_0} \iint [\mathbf{B} \cdot (\nabla \times \mathbf{A}) - \nabla \cdot (\mathbf{A} \times \mathbf{B})] d\Omega = \frac{1}{2\mu_0} \iint [\mathbf{B} \cdot \mathbf{B} - \nabla \cdot (\mathbf{A} \times \mathbf{B})] d\Omega$$

$$= \frac{1}{2\mu_0} \left[\iint B^2 d\Omega - \oint_S (\mathbf{A} \times \mathbf{B}) \cdot d\mathbf{a} \right] \tag{2.58}$$

Here divergence theorem is used for the second term. If we take the limit of volume integral large enough, both $\mathbf{A}$ and $\mathbf{B}$ become negligibly small and the surface integral term vanishes. Thus, we can find the remaining integrand of Eq. (2.58) as the magnetic energy density w_m.

$$w_m = \frac{1}{2\mu_0} B^2 \tag{2.59}$$

Together with Eq. (2.48), we find the total electromagnetic energy in free space as follows:

$$w_{em} = w_e + w_m = \frac{1}{2} \epsilon_0 E^2 + \frac{1}{2\mu_0} B^2 \tag{2.60}$$

Later in this book, we will see that a light wave carries w_{em} at the speed of light. Here in Eq. (2.60) the electric permittivity ϵ_0 and magnetic permeability μ_0 are those for free space (vacuum). When a light wave propagates through a medium, these quantities must be replaced with ϵ and μ, respectively, as they are medium-dependent constants (although $\mu \cong \mu_0$ for most non-magnetic materials).

2.2.6   Algebraic Explanation of Electromagnetic Waves

Faraday's law (2.3) and Ampère's law (2.2), respectively, tell us that the temporal change of $\mathbf{B}$ is a source for curl of $\mathbf{E}$ and the temporal change of $\mathbf{E}$ is a source for the curl of $\mathbf{B}$. By substituting one of these equations into the other, we can eliminate one of the field vectors and derive the wave equation that represents the spatiotemporal behavior of the remaining field.

In deriving the wave equation, we drop the conduction current term from the right-hand side of Eq. (2.2). As mentioned above, the other term (Maxwell's term), together with the right-hand side of Faraday's law, generates the wave dynamics of the electromagnetic field. The conduction current term causes energy dissipation, which attenuates the electromagnetic wave. After dropping the conduction current term, we differentiate both-hand sides of Eq. (2.2) with respect to time. At the same time, we take the curl of Eq. (2.3) using the mathematical identity $\nabla \times \nabla \times \mathbf{E} = \nabla(\nabla \cdot \mathbf{E}) - \nabla^2\mathbf{E}$.

$$\frac{\partial(\nabla \times \mathbf{B})}{\partial t} = \epsilon\mu\frac{\partial^2\mathbf{E}}{\partial t^2} \tag{2.61}$$

$$\nabla^2\mathbf{E} = \frac{\partial(\nabla \times \mathbf{B})}{\partial t} \tag{2.62}$$

Here in deriving Eq. (2.62), we set $\nabla \cdot \mathbf{E} = 0$ because we assume that there is no electric charge in the space. From Eqs. (2.61) and (2.62), we can eliminate $\mathbf{B}$. Using $c = 1/\sqrt{\epsilon\mu}$ for the phase velocity of light, we derive the following wave equation for the electric field:

$$\nabla^2\mathbf{E} = \frac{1}{c^2}\frac{\partial^2\mathbf{E}}{\partial t^2} \tag{2.63}$$

Repeating a similar procedure, we can eliminate $\mathbf{E}$ and obtain the following wave equation for the magnetic field.

$$\nabla^2\mathbf{B} = \frac{1}{c^2}\frac{\partial^2\mathbf{B}}{\partial t^2} \tag{2.64}$$

Equations (2.63) and (2.64) are wave equations for electromagnetic waves that propagate in a lossless medium (e.g., vacuum).

When an electromagnetic wave propagates through a conductive medium, we need to keep the conduction current term $\mu\mathbf{j}$ in Maxwell's equation (2.2). Using Ohm's law we can put the wave equation (2.63) in the following form: (See p. 392–395 of [2] for the wave equation and solution when the space contains charges.)

$$\nabla^2\mathbf{E} = \frac{1}{c^2}\frac{\partial^2\mathbf{E}}{\partial t^2} + \mu\sigma\frac{\partial\mathbf{E}}{\partial t} \tag{2.65}$$

Here σ is the conductivity. The differential equation (2.65) has the term containing the first-order time derivative $\partial\mathbf{E}/\partial t$. As we observed in the equation of motion (1.4) that describes decaying spring oscillation, this term causes exponential decay. We will discuss this topic in the next section.

2.2.7   Plane Wave Solutions

Electromagnetic waves propagating in vacuum can be found as solutions to differential equations (2.63) and (2.64). There are a number of methods [21] to solve this type of differential equations. Here, we consider a simple way. We know that the solution has temporal and spatial periodicity and the phase term contains the information. Let ω and k represent the temporal and spatial angular frequencies, respectively. Then we can put the phase of the solution in the following form:

$$\theta = \omega t \pm kz \tag{2.66}$$

Using $E(t, \mathbf{r}) = f(t, \mathbf{r})$ to denote the solution to Eq. (2.63), we can immediately find the following equality:

$$k^2 f'' = \frac{\omega^2}{c^2} f'' \tag{2.67}$$

Here f'' denotes the second-order derivative of function f. Since $f'' \neq 0$ (because if $f'' = 0$, f is a quadratic function and would not be periodic), Eq. (2.67) leads to the following equation:

$$c^2 = \frac{\omega^2}{k^2} \tag{2.68}$$

Equation (2.68) indicates that the wave velocity c is the ratio of the temporal periodicity ω to the spatial periodicity k. As we have discussed in Chap. 1, the wave velocity in this form is the phase velocity and a plane wave solution satisfies the wave equation in the form of (2.63). So, we can put the simplest form of solution as follows:

$$E(t, \mathbf{r}) = E_0 \cos(\omega t \pm \mathbf{k} \cdot \mathbf{r}) \tag{2.69}$$

It is obvious that we can repeat the same process to find the following form of solution for **B**.

$$B(t, \mathbf{r}) = B_0 \cos(\omega t \pm \mathbf{k} \cdot \mathbf{r}) \tag{2.70}$$

At the end of Sect. 2.1, from dimensional analysis of Faraday's law, we found that the amplitude of the electric field of a light wave is the speed of light times higher than the amplitude of the magnetic field. Now, we are in a position to confirm it. Substitute Eq. (2.69) into Faraday's law (2.3). For simplicity, set our x-axis in line with the electric field and the light wave travels along the z-axis. We do not lose the generality by doing so because at a given point the electric field of a light wave is oriented in a certain direction and it can be our x-axis and as a transverse wave the light travels in a direction perpendicular to the

x-axis and it happens to be our z-direction. With this coordinate axis, Eq. (2.69) becomes

$$\mathbf{E}(t, \mathbf{r}) = E_0 \cos(\omega t \pm k_z z)\hat{x} \qquad (2.71)$$

Substitution of Eq. (2.71) into the left-hand side of Eq. (2.3) yields the following equation:

$$\nabla \times \mathbf{E}(t, \mathbf{r}) = \frac{\partial E_x}{\partial z}\hat{y} = \mp k_z E_0 \sin(\omega t \pm k_z z)\hat{y} \qquad (2.72)$$

Equation (2.72) and Faraday's law tell us that the magnetic field is polarized along the y-axis for the following reason. The right-hand side of Faraday's law is the temporal derivative of the magnetic field. Since the temporal differentiation does not change the spatial orientation of the vector, this indicates that the magnetic field has the same orientation as the left-hand side of Faraday's law. Equation (2.72) tells us in this case the left-hand side of Faraday's law represents a vector oriented along the y-axis. This argument allows us to put the magnetic field in the following form:

$$\mathbf{B}(t, \mathbf{r}) = B_0 \cos(\omega t \pm k_z z)\hat{y} \qquad (2.73)$$

Further, the substitution of Eq. (2.73) into the right-hand side of Eq. (2.3) yields the following form of the magnetic field vector:

$$-\frac{\partial \mathbf{B}(t, \mathbf{r})}{\partial t} = \omega B_0 \sin(\omega t \pm k_z z)\hat{y} \qquad (2.74)$$

From Eq. (2.3), the amplitudes of Eq. (2.72) and (2.74) must be the same. This leads to the following relation:

$$E_0 = \frac{\omega}{k_z} B_0 = c B_0 \qquad (2.75)$$

Equation (2.75) states that the amplitude of the electric field is the speed of light times the amplitude of the magnetic field.

In fact, this statement has some subtleness that indicates the relativistic nature of the electromagnetic field and Maxwell's theory. Rearrangement of Eq. (2.75) using the wavelength λ and period τ into $E_0/\lambda = B_0/\tau$ tells us that the division of the amplitude of the electric field vector of light by the spatial periodicity is equal to the division of the amplitude of the magnetic field vector by the temporal periodicity. This means that if we count the electric field with the spatial unit of light the result is the same as counting the magnetic field with the temporal unit of light. This naively indicates the space–time relation [22, 23]. Clearly, the connection between light and space–time is profound.

When the electromagnetic wave propagates through a medium that contains electric charges, we need to solve the wave equation (2.65). The solution takes the following form, which has the exponentially decaying term, as we expected.

$$\mathbf{E}(t, \mathbf{r}) = E_0 e^{-\kappa t} \cos(\omega t \pm k_z z)\hat{x} \tag{2.76}$$

$$\mathbf{B}(t, \mathbf{r}) = B_0 e^{-\kappa t} \cos(\omega t \pm k_z z)\hat{y} \tag{2.77}$$

Here the decay constant κ is given as follows:

$$\kappa = \omega \sqrt{\frac{\epsilon\mu}{2}} \left[\sqrt{1 + \left(\frac{\sigma}{\epsilon\omega}\right)^2} - 1 \right]^{1/2} \tag{2.78}$$

The solutions discussed in this section are referred to as the plane wave solutions [24] because the wavefront is a plane. The wavefront is defined as the locus of a constant phase. The wave solutions (2.69), (2.70), (2.76), and (2.77) all have the phase term independent of x and y, the coordinate variables perpendicular to the axis of propagation. Thus, the wavefront is planar. In addition, the amplitude is independent of x and y as well. The waves having this property are not realistic because the amplitude represents the energy (as we will discuss in the next section) and if the energy is constant in a plane perpendicular to the axis of propagation the wave carries infinite energy. However, in an infinitesimal area around the axis of propagation, we can always find a tangential plane. Therefore, the argument regarding the propagation of a plane wave is generally correct in the vicinity of the propagation axis. Later we will discuss the propagation of Gaussian beams. There the on-axis beam behaves like a plane wave.

2.2.8 Light Wave as a Flow of Electromagnetic Energy

In Chap. 1, we discussed that every wave carries energy. It is clear that a light wave carries electromagnetic energy. We also discussed that a wave consists of a force-like component and a velocity-like component. When we discussed sound waves in Sect. 1.1.4, the pressure wave was the force-like component and the particle velocity wave was the velocity-like component. An electromagnetic wave consists of an electric wave component and a magnetic component. These arguments naturally lead to the interpretation that one of the fields behaves like force and the other behaves like velocity. Which field behaves like force and which behaves like velocity? You will probably correctly guess that the electric wave is the force-like component and the magnetic wave is the velocity-like component. The reason is as follows.

When a light wave propagates through a medium, the electric field exerts an electric force on the charges. As a result, the charges move to form a current, which in turn generates a magnetic field. It is analogous to the case in elastic waves where the stress wave pushes atoms, and consequently the atoms gain velocity. It is natural to interpret that the electric wave is force-like and the magnetic wave is velocity-like.

Now that we identify the electric and magnetic fields as representing force-like and velocity-like quantities, we can continue the discussion by viewing their product as a flow of energy. Light is a transverse wave where the electric and magnetic field components are orthogonal to each other. The most natural way to interpret it as an energy flow is to consider the vector product of the electric and magnetic field vectors. The amplitude of the resultant vector is the product of force and velocity, and its direction is perpendicular to both the electric and magnetic fields. The energy vector of this nature is known as the Poynting vector $\mathbf{S}$, which we can argue from the law of energy conservation [25]. The associated theorem is known as Poynting's theorem.

Poynting's theorem can be expressed as follows:

$$\frac{\partial u}{\partial t} = -\nabla \cdot \mathbf{S} - \mathbf{j} \cdot \mathbf{E} \tag{2.79}$$

Here u is the electromagnetic energy density, $\mathbf{j}$ is the current density, $\mathbf{E}$ is the electric field, and $\mathbf{S}$ is the Poynting vector defined as follows:

$$\mathbf{S} = \frac{\mathbf{E} \times \mathbf{B}}{\mu} \tag{2.80}$$

In Eq. 2.80, B is the magnetic field and μ is the magnetic permeability of the medium through which the light wave propagates.

We can view Eq. (2.79) as a divergence theorem or an equation of continuity that indicates the electromagnetic energy is a conserved quantity. The left-hand side is the temporal change of the electromagnetic energy density. On the right-hand side, the first term is the energy flow out of the volume per unit time and the second term is the work done by the electric field to move the electric charges. The first term represents the power of the electromagnetic wave coming out of the unit volume. The second term represents the power that the electric field loses. The entire equation indicates that the change in the electromagnetic energy over time is equal to the sum of the power of the electromagnetic wave traveling out of the volume and the power dissipation due to the conduction current.

It is interesting to express the amplitude of the Poynting vector, S_0, using the electric field or magnetic field's amplitude only. Using that the ratio of the amplitude of the electric field over the amplitude of the magnetic field is the speed of light (Eq. (2.75)), and we obtain the following expressions:

$$S_0 = \frac{E_0 B_0}{\mu} = \frac{E_0 (E_0/c)}{\mu} = \frac{E_0^2}{c\mu} \tag{2.81}$$

$$S_0 = \frac{E_0 B_0}{\mu} = \frac{(B_0 c) B_0}{\mu} = \frac{c B_0^2}{\mu} \tag{2.82}$$

Here c is the speed of light.

Since we know that the electromagnetic energy flux travels at the speed of light (see Eqs. (3.64) and (3.65)) that indicate that the Poynting vector propagates at the same velocity as the electric and magnetic waves), we can express the amplitude of the $\mathbf{S}$ wave as $S_0 = uc$. Hence, using Eq. (2.81) , we obtain the following expression for u.

$$u = \frac{S_0}{c} = \frac{E_0^2}{c^2 \mu} = \epsilon E_0^2(x, y) \tag{2.83}$$

Here, $c = 1/\sqrt{\epsilon \mu}$. Note that u in this form is twice of the electric energy density $\epsilon E_0^2/2$ defined by Eq. (2.48).

Similarly, using Eq. (2.82), we obtain the following expression of u:

$$u = \frac{S_0}{c} = \frac{B_0^2(x, y)}{\mu} \tag{2.84}$$

In this form, u is twice the magnetic energy density $B_0^2/(2\mu)$ defined by Eq. (2.59).

Equations (2.83) and (2.84) indicate that as a light wave propagates, the electromagnetic energy stored in the space flows at the speed of light. Recall in Sect. 2.2.3, we discussed that the energy is stored both in the capacitor and inductor, and they alternate in time with the same maximum energy. Equation (2.60) indicates the total electromagnetic energy is the sum of the capacitive and inductive energies. Both (2.83) and (2.84) represent the total energy in the unit volume. The former expresses the total energy in terms of the electric field amplitude E_0, and the latter of the magnetic field amplitude B_0. Since the maximum electric energy is the same as the maximum magnetic energy, the total energy is double of maximum electric or magnetic energy.

References

1. P. G. Huray, *Maxwell's equations*, (IEEE Press, Wiley, Hoboken, USA, 2010)
2. D. J. Griffiths, *Introduction to electrodynamics*. 3rd edn. (Prentice Hall, Upper Saddle River, NJ, USA, 1999)
3. A. Yariv, *Introduction to Optical Electronics*, (Holt, Rinehart and Winston, New York, USA, 1971)
4. J. C. Slater, N. H. Frank, *Electromagnetism*, (McGraw-Hill, New York, USA, 1947)
5. The Divergence Theorem, https://openstax.org/books/calculus-volume-3/pages/6-8-the-divergence-theorem (accessed on August 4, 2022)
6. Electric Potential, openstax, https://openstax.org/books/physics/pages/18-4-electric-potential (accessed on August 4, 2022)
7. K. D. Bonin, V. V. Kresin, *Electric-dipole Polarizabilities of atoms, molecules, and clusters* (World Scientific, Singapore, New Jersey, 1997)
8. H. Jakubowski, D2. Photoexcitation and Electron Transfer, https://chem.libretexts.org/Bookshelves/Biological_Chemistry/Book%3A_Biochemistry_Online_(Jakubowski)/10%3A_Oxidation/10.4%3A_Photosynthesis_-_The_Light_Reaction/D2._Photoexcitation_and_Electron_Transfer (accessed on August 4, 2022)
9. Stokes' Theorem, openstax, https://openstax.org/books/calculus-volume-3/pages/6-7-stokes-theorem
10. D. J. Griffiths, *Introduction to electrodynamics*. 3rd edn. (Prentice Hall, Upper Saddle River, NJ, USA, 1999), pp. 322-324
11. A. Giambattista, B. M. Richardson, R. C. Richardson *College physics.*, (Mc Graw Hill, NY, USA, 2004), pp. 798

12. M. Bezerra, W. Kort-Kamp, M. V. Cougo-Pinto, C. Farina de Souza, How to introduce the magnetic dipole moment, Euro. J. Phys. **33** (5) 1313-1320, 2012
13. D. J. Griffiths, *Introduction to electrodynamics*. 3rd edn. (Prentice Hall, Upper Saddle River, NJ, USA, 1999), pp. 243
14. N. Forbes, B. Mahon, *Faraday, Maxwell, and the Electromagnetic Field, How two men revolutionized physics* (Prometheus Books, New York, 2014)
15. Lenz Law vs. Faraday's Law, https://resources.pcb.cadence.com/blog/2020-lenz-law-vs-faradays-law-how-do-they-govern-crosstalk-and-emi (accessed on August 4, 2022)
16. Oscillations in an LC Circuit, openstax, https://openstax.org/books/university-physics-volume-2/pages/14-5-oscillations-in-an-lc-circuit
17. P. C. Magnusson, G. C. Alexander, V. K. Tripathi, A Weisshaar, *Transmission Lines and Wave Propagation*, 4th edn. (CRC Press, Boca Raton, London, 2001)
18. Capacitors and Dielectrics, openstax, https://openstax.org/books/physics/pages/18-5-capacitors-and-dielectrics (accessed on August 4, 2022)
19. J. Coopersmith, The Principle of Virtual Work in *The Lazy Universe: An Introduction to the Principle of Least Action* (Oxford Univ. Press., 2017), pp. 59–87. https://doi.org/10.1093/oso/9780198743040.003.0004
20. K. Ó. Klausen, *A Treatise on the Magnetic Vector Potential*, (Springer Nature Switzerland AG, Cham, Switzerland, 2020)
21. Second-Order Linear Equations, LibreTexts, https://math.libretexts.org/Bookshelves/Calculus/Book%3A_Calculus_(OpenStax)/17%3A_Second-Order_Differential_Equations/17.1%3A_Second-Order_Linear_Equations
22. A. Einstein, *The meaning of Relativity*, 5th edn., (MJF Books, New York, 1984)
23. D. J. Griffiths, *Introduction to electrodynamics*. 3rd edn. (Prentice Hall, Upper Saddle River, NJ, USA, 1999), pp. 522–535
24. C. J. Papachristou, Plane-wave solutions of Maxwell's equations: An educational note, https://arxiv.org/ftp/arxiv/papers/2003/2003.06510.pdf (accessed on August 4, 2022)
25. J. C. Slater, N. H. Frank, *Electromagnetism*, (McGraw-Hill, New York, USA, 1947), chap. 8

Light Propagation in Matter

3

3.1 Maxwell Equations in Matter

3.1.1 Electric Dipole and Polarization

An electric dipole [1] consists of a pair of unlike electric charges separated by a small distance.[1] We can define the electric dipole moment vector as $\mathbf{p} = q\mathbf{d}$ where q is the electric charge and $\mathbf{d}$ is the position vector pointing from the negative to the positive charge of the dipole (Fig. 3.1).

A dielectric material [2] is an insulator, meaning that it does not have free electrons. Instead, it has electrons bound to the nucleus. When an electric field is applied, the nuclei (the positive charges) are pushed and the bound electrons (the negative charges) are pulled by the electric force. Consequently, the dipole moment [3, 4] is induced.

$$\mathbf{p} = \alpha\mathbf{E} \qquad (3.1)$$

The constant of proportionality α is called the atomic polarizability[4]; the greater the α, the greater the d. Note that $\mathbf{p}$ is parallel to $\mathbf{E}$.

The polarization [5] P is defined as the dipole moment per unit volume. Figure 3.2 illustrates schematically polarization induced in a dielectric material by an external electric field.

Let q be the total charge piled up at the right end of the dielectric material. The cross-sectional area is S. Since the volume is Sd, the total dipole moment of the volume is PSd. This must be equal to the total dipole moment qd.

$$qd = \rho_b(Sd) = PSd \qquad (3.2)$$

---

[1] How close? Close enough so that the oscillatory effects discussed in this section are effective.

© The Author(s), under exclusive license to Springer Nature Switzerland AG 2023
S. Yoshida, *Fundamentals of Optical Waves and Lasers*, Synthesis Lectures on Wave Phenomena in the Physical Sciences. https://doi.org/10.1007/978-3-031-18188-7 3

Fig. 3.1 Electric dipole

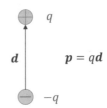

Fig. 3.2 Polarization induced
in the dielectric. q represents
the positive charge in a unit
volume on the right surface of
the dielectric

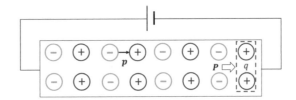

Here, ρ_b is the bound charge density. From the left- and right-hand sides of Eq. (3.2), we
find the following equation:

$$q = PS \tag{3.3}$$

By dividing Eq. (3.3) by the cross-sectional area S, we find that the polarization is equal to
the surface charge density of the bound charge σ_b.

$$P = \frac{q}{S} \equiv \sigma_b \tag{3.4}$$

Interpreting the total charge q as the volume integration of the bound charge density ρ_b, we
obtain the following expression:

$$q = \int_V \rho_b dV \tag{3.5}$$

Referring to Eq. (3.4), we can view σ_b as the magnitude of the polarization vector. Consider
the area integration over the positive charge layer at the right surface (enclosed by a dashed
rectangle) in Fig. 3.2. Since the vector $\mathbf{P}$ is toward (into) this layer on the surface and there is
no polarization vector on the other surfaces of this layer, we can express the area integration
as follows:

$$\oint_S \mathbf{P} \cdot d\mathbf{a} = \int_{\text{left}} P(\hat{\mathbf{i}}) \cdot da(-\hat{\mathbf{i}}) = -P \int_{\text{left}} da = -PS = -\frac{q}{S}S = -q \tag{3.6}$$

Here, we used Eq. (3.4) in the last step of Eq. (3.6).

From Eqs. (3.5) and (3.6), we find the following equation:

$$\oint_S \mathbf{P} \cdot d\mathbf{a} = -q = -\int_V \rho_b dV \tag{3.7}$$

Applying the divergence theorem [7], we can write the left-hand side of Eq. (3.6) as follows:

$$\oint_S \mathbf{P} \cdot d\mathbf{a} = \int_V \nabla \cdot \mathbf{P} dV \tag{3.8}$$

From Eqs. (3.7) and (3.8), we derive the following equation:

$$-\int_V \rho_b dV = \int_V \nabla \cdot \mathbf{P} dV \tag{3.9}$$

Since Eq. (3.9) holds for any volume, we can relate the Polarization vector with the bound charge density as follows [5, 6]:

$$\nabla \cdot \mathbf{P} = -\rho_b \tag{3.10}$$

Remember that Gauss's law (2.1) states the total charge density is the divergence of the electric field times the electric permittivity of free space.

$$\epsilon_0 \nabla \cdot \mathbf{E} = \rho \tag{3.11}$$

We can express the total charge density ρ as the sum of the bound charge density ρ_b and free (unbound) charge density ρ_f.

$$\rho = \rho_b + \rho_f \tag{3.12}$$

From Eqs. (3.10), (3.11) and (3.12), we find the following equation:

$$\epsilon_0 \nabla \cdot \mathbf{E} = \rho = \rho_b + \rho_f = -\nabla \cdot \mathbf{P} + \rho_f \tag{3.13}$$

From Eq. (3.13), we derive the following expression for ρ_f:

$$\nabla \cdot (\epsilon_0 \mathbf{E} + \mathbf{P}) = \rho_f \tag{3.14}$$

Defining the electric flux density vector (also known as the electric displacement vector) $\mathbf{D}$ as follows:

$$\mathbf{D} = \epsilon_0 \mathbf{E} + \mathbf{P} \tag{3.15}$$

we can write Eq. (3.11) in the following form:

$$\nabla \cdot \mathbf{D} = \rho_f \tag{3.16}$$

Equation (3.16) is Gauss's law in matter corresponding to Gauss' law in free space (2.1).

Similarly to the electric field, we can introduce the auxiliary magnetic field $\mathbf{H}$ as follows [8]:

$$\mathbf{H} = \frac{1}{\mu_0} \mathbf{B} - \mathbf{M} \tag{3.17}$$

Here, $\mathbf{M}$ is the magnetization vector. Magnetization is defined as the magnetic dipoles per unit volume, the magnetic version of the polarization P.

Using **H**, we can write Ampère's law in the following form:

$$\nabla \times \mathbf{H} = \frac{\partial \mathbf{D}}{\partial t} + \mathbf{j}_f \tag{3.18}$$

Here, $\mathbf{j}_f$ is the free current density. Derivation of Eq. (3.18) and the physical meanings of the auxiliary field **H** and free current are analogous to the electric version of them, i.e., Eq. (3.16), **D** and ρ_f but a little more lengthy. We will not discuss them in this book. Chapters 6 and 7 of [9] discuss these topics in detail.

3.1.2   Linear Medium

In a linear medium [6, 10], the electric flux density vector and auxiliary magnetic field vector are proportional to the electric field and magnetic field, respectively.

$$\mathbf{D} = \epsilon \mathbf{E} \tag{3.19}$$

$$\mathbf{H} = \frac{1}{\mu} \mathbf{B} \tag{3.20}$$

The constants of proportionality ϵ and μ are the electric permittivity and magnetic permeability of the linear material.

Substitutions of Eq. (3.15) and (3.17) into the left-hand sides of Eqs. (3.19) and (3.20) leads to the following equations:

$$\epsilon_0 \mathbf{E} + \mathbf{P} = \epsilon \mathbf{E} \tag{3.21}$$

$$\frac{1}{\mu_0} \mathbf{B} - \mathbf{M} = \frac{1}{\mu} \mathbf{B} \tag{3.22}$$

Equations (3.21) and (3.22) indicate that in a linear medium the polarization is proportional to the applied field. In other words, the polarization increases in proportion to the applied field. Later in this chapter, we introduce the electric susceptibility χ_e as $\mathbf{P} = \epsilon_0 \chi_e \mathbf{E}$. The linearity means that the electric susceptibility is a constant. If it is dependent on the electric field, the polarization has a quadratic or higher order dependence on the electric field. Under the application of an alternating electric field, e.g., a light wave, nonlinear materials generate multiple frequencies. Although we will not discuss the topic, this multiple-frequency generation with nonlinear material is important in numerous applications. The reader is encouraged to consult the references [11, 12]. (Also see Chap. 8 (p. 258–308) of [13].)

3.2   Light–Matter Interaction

In Sect. 2.1, we considered an electron orbiting around the nucleus as an example of an electric dipole. The electron is bound because the electric force exerted by the nucleus acts as the centripetal force that maintains the orbiting motion of the electron. When this electric dipole is placed in an alternating electric field (external field), the electron is pulled toward the nucleus when the external electric field vector is opposite to $\mathbf{d}$ and is pushed further out from the orbit when the external electric field is in the same direction as $\mathbf{d}$. Since the nucleus is much heavier than the electron, we can assume that the nucleus is motionless, while the electron is pulled or pushed by the external electric field.

We can view the dynamics as a spring–mass system with an external alternating force driving the mass. Applying the argument of the damped driven harmonic oscillator, we made in Sect. 1.1, we can express the motion of the electron by the following equation of motion [2]:

$$\frac{d^2\tilde{x}}{dt^2} + \gamma\frac{d\tilde{x}}{dt} + \omega_0^2\tilde{x} = \frac{qE_0}{m}e^{i\omega t} \tag{3.23}$$

Here, $\tilde{x}$ is the complex displacement of the electron, γ and ω_0, respectively, denote the overall damping effect and the natural angular frequency of the electron, m and q are the mass and charge of the electron, E_0 and ω are the amplitude and angular frequency of the external field.

In the steady state, the electron oscillates at the driving frequency ω. (Remember the discussion made in association with the particular solution in Chap. 1) Therefore, we can put the solution $\tilde{x}$ as follows:

$$\tilde{x} = \tilde{x}_0 e^{i\omega t} \tag{3.24}$$

Here, $\tilde{x}_0$ is the complex amplitude of the displacement. By substituting Eq. (3.24) into Eq. (3.23), we find

$$\tilde{x}_0 = \frac{q/m}{\omega_0^2 - \omega^2 - i\gamma\omega}E_0 \tag{3.25}$$

Thus, the solution to Eq. (3.23) in general, takes the following form:

$$\tilde{x}(t) = \frac{q/m}{\omega_0^2 - \omega^2 - i\gamma\omega}E_0 e^{i\omega t} \tag{3.26}$$

Since $\tilde{x}(t)$ represents the distance between the nucleus and electron, the dipole can be expressed as the product of this quantity and the charge q:

$$\tilde{p}(t) = q\tilde{x} = \frac{q^2/m}{\omega_0^2 - \omega^2 - i\gamma\omega}E_0 e^{i\omega t} \tag{3.27}$$

The real part of the quantity $\tilde{p}(t)$ is the dipole moment.

---

[2] Sect. 5.4 of [13] explains the content of this section in more detail.

Each atom (or molecule) has a number of electrons that have different natural frequencies and damping coefficients. In a unit volume, there are a number of atoms. We can express the overall effect of all dipoles in the volume by adding the dipole moment. The real part of the resultant quantity, $\tilde{P}$, is called the polarization $\mathbf{P}$. If there are f_j atoms in the unit volume, we can write $\tilde{P}$ as follows:

$$\tilde{\mathbf{P}} = \sum \tilde{p}_j f_j = \frac{Nq^2}{m} \left(\sum_j \frac{f_j}{\omega_{j0}^2 - \omega^2 - i\gamma_j\omega} \right) \tilde{\mathbf{E}} \tag{3.28}$$

Here, N is the total number of atoms in the unit volume, suffix j identifies the electron whose natural angular frequency is ω_{j0}, damping coefficient is γ_j, and $\tilde{E} \equiv E_0 exp(-i\omega t)$ is the complex representation of the external electric field.

As we discussed in Sect. 1.1, the solution (3.26) represents a harmonic oscillation whose amplitude depends on the driving frequency (ω) relative to the natural frequency (ω_0), and the decay constant γ. We discussed that the oscillation amplitude is high when the driving frequency is the same or near the natural frequency and the decay constant is low. When the driving frequency is equal to the natural frequency, we say that a resonant condition is established. In the present context, it is important to note that since an atomic system is a quantum system the electron takes discrete energy levels. In other words, unlike the spring–mass system discussed in Sect. 1.1, the radius (the distance between the electron and nucleus) can take only a discrete value. Therefore, in the present context, it makes more sense to discuss the amplitude and the phase of the polarization P in terms of the interaction of the atom with the electric field of the light incident to the atom. Thus, we introduce the electric susceptibility $\tilde{\chi}_e$ as follows:

$$\tilde{P} = \epsilon_0 \tilde{\chi}_e \tilde{E} \tag{3.29}$$

From the comparison of Eqs. (3.28) and (3.29), we find

$$\tilde{\chi}_e = \frac{Nq^2}{m\epsilon_0} \left(\sum_j \frac{f_j}{\omega_{j0}^2 - \omega^2 - i\gamma_j\omega} \right) \tag{3.30}$$

Substitution of Eq. (3.29) into Eq. (3.15) yields the following expression:

$$\tilde{D} = \epsilon_0 \tilde{E} + \epsilon_0 \tilde{\chi}_e \tilde{E} = \epsilon_0 (1 + \tilde{\chi}_e) \tilde{E} \tag{3.31}$$

Equation (3.31) indicates the proportionality between $\tilde{D}$ and $\tilde{E}$. The constant of this proportionality is referred to as the complex permittivity of the dielectric material.

$$\tilde{D} = \epsilon_0 (1 + \tilde{\chi}_e) \tilde{E} \equiv \tilde{\epsilon} \tilde{E} \tag{3.32}$$

Sometimes, the following expressions are used. (The subscripts or superscripts r and i denote *real* and *imaginary*.)

$$\tilde{\epsilon}_r = 1 + (\tilde{\chi}_e)^r, \quad \tilde{\epsilon}_i = (\tilde{\chi}_e)^i \tag{3.33}$$

Equation (3.32) indicates that through the electric susceptibility, the dipole changes the electric permittivity ϵ_0 to $\tilde{\epsilon}$. This is the interaction between the incident light and the atom. Due to the electric field of the incident light, the electric permittivity inside the medium is altered as indicated by Eq. (3.32).

3.2.1   Absorption of Light and Index of Refraction

Now consider the interaction of light and matter from the viewpoint of light. For simplicity, we consider a plane wave but the gist of the argument is the same for non-plane waves. We will discuss Gaussian beams, non-planar, and realistic laser beams, in Sect. 3.3.

Consider the following wave equation and plane wave solution:

$$\frac{\partial^2 \tilde{E}}{\partial t^2} - \frac{1}{\tilde{\epsilon}\mu_0}\nabla^2 \tilde{E} = 0 \tag{3.34}$$

$$\tilde{E}(z, t) = \tilde{E}_0 e^{i(\omega t - \tilde{k}z)} \tag{3.35}$$

Here, $\tilde{k}$ is the complex wave number.

$$\tilde{k} = k_r + ik_i \tag{3.36}$$

The imaginary part represents energy dissipation in the medium. Substitution of solution (3.35) into wave equation (3.34) leads to the following expression of $\tilde{k}$:

$$\tilde{k} = \omega\sqrt{\tilde{\epsilon}\mu_0} = \omega \left(\epsilon_0(1 + \tilde{\chi}_e)\mu_0\right)^{1/2} \cong \omega\sqrt{\epsilon_0\mu_0}\left(1 + \frac{1}{2}\tilde{\chi}_e\right) \tag{3.37}$$

$$= \frac{\omega}{c_0}\left(1 + \frac{1}{2}\tilde{\chi}_e\right) = \frac{\omega}{c_0}\left[1 + \frac{Nq^2}{2m\epsilon_0}\left(\sum_j \frac{f_j}{\omega_{j0}^2 - \omega^2 - i\gamma_j\omega}\right)\right] \tag{3.38}$$

Here, we used Eqs. (3.30), (3.32), $c_0 = 1/\sqrt{\epsilon_0\mu_0}$ (Eq. (2.25) applied to vacuum), and binomial approximation $\sqrt{1 + \tilde{\chi}_e} \approx 1 + \tilde{\chi}_e/2$ assuming $1 >> \tilde{\chi}_e$.

By separating the right-hand side of Eq. (3.38) into the real and imaginary parts, we can clarify the physical meanings of n as the phase part of the interaction between the driving electric field (light) and the dipole.

$$\tilde{k} = \frac{\omega}{c_0} \left[1 + \frac{Nq^2}{2m\epsilon_0} \left(\sum_j \frac{f_j(\omega_{j0}^2 - \omega^2)}{(\omega_{j0}^2 - \omega^2)^2 + \gamma_j^2\omega^2} \right) + i \frac{Nq^2}{2m\epsilon_0} \left(\sum_j \frac{f_j\gamma_j\omega}{(\omega_{j0}^2 - \omega^2)^2 + \gamma_j^2\omega^2} \right) \right]$$

(3.39)

Therefore,

$$k_r = \frac{\omega}{c_0} \left[1 + \frac{Nq^2}{2m\epsilon_0} \left(\sum_j \frac{f_j(\omega_{j0}^2 - \omega^2)}{(\omega_{j0}^2 - \omega^2)^2 + \gamma_j^2\omega^2} \right) \right]$$

(3.40)

$$k_i = \frac{Nq^2}{2m\epsilon_0 c_0} \left(\sum_j \frac{f_j\gamma_j\omega^2}{(\omega_{j0}^2 - \omega^2)^2 + \gamma_j^2\omega^2} \right)$$

(3.41)

The index of refraction n is c_0/c, where c_0 is the speed of light in free space (vacuum) and c is the speed of light in the medium. Since the wave velocity is the ratio of the temporal frequency over the spatial frequency

$$c = \frac{\omega}{k_r},$$

(3.42)

n can be expressed as follows:

$$n = \frac{c_0}{c} = \frac{c_0}{\omega/k_r} = \frac{c_0}{\omega} k_r$$

(3.43)

Substituting Eq. (3.40) into Eq. (3.43), we obtain the following expressions for the index of refraction and absorption coefficient α:

$$n = \frac{c_0}{c} = \frac{c_0}{\omega} k_r = 1 + \frac{Nq^2}{2m\epsilon_0} \left(\sum_j \frac{f_j(\omega_{j0}^2 - \omega^2)}{(\omega_{j0}^2 - \omega^2)^2 + \gamma_j^2\omega^2} \right)$$

(3.44)

$$\alpha = 2k_i = \frac{Nq^2}{m\epsilon_0 c_0} \left(\sum_j \frac{f_j\gamma_j\omega^2}{(\omega_{j0}^2 - \omega^2)^2 + \gamma_j^2\omega^2} \right)$$

(3.45)

Note that the quantity in the parenthesis of Eq. (3.44) is half of the real part of the complex electric susceptibility $\tilde{\chi}_e$ defined in Eq. (3.30), indicating that n is due to the susceptibility. We can interpret Eq. (3.44) as follows: "the speed of light in the dielectric medium decreases in proportion to the extent the electric field interacts with the dipoles via $\tilde{\chi}_e$". Equation (3.45) indicates that "the absorption of a light wave is due to the damping effect of the dielectric medium". Also, note that α is the power absorption coefficient. The factor 2 in front of k_i comes from the fact that the power P is proportional to E^2 hence $P \propto (e^{-k_i z})^2 = e^{-2k_i z} = e^{-\alpha z}$. The frequency dependence of the index of refraction seen in Eq. (3.44) means that the wave velocity depends on the frequency; hence, it represents dispersion (See Sect. 4.3).

Let's consider the complex electric susceptibility in conjunction with light absorption. Absorption occurs when the absorbing species' dipole moment and the absorbed light are in resonance. Therefore, we can approximate the light's frequency by the natural frequency, $\omega \cong \omega_0$. Using this approximation, we can write Eq. (3.30) as follows:

$$\tilde{\chi}_{abs} = \frac{Nq^2}{m\epsilon_0}\left(\frac{1}{\omega_0^2 - \omega^2 - i\gamma\omega}\right) = \frac{Nq^2}{m\epsilon_0}\left(\frac{1}{(\omega_0 + \omega)(\omega_0 - \omega) - i\gamma_j\omega}\right)$$

$$\cong \frac{Nq^2}{m\epsilon_0}\left(\frac{1}{2\omega_0(\omega_0 - \omega) - i\gamma\omega_0}\right) = \frac{Nq^2}{m\epsilon_0\omega_0}\left(\frac{1}{2(\omega_0 - \omega) - i\gamma}\right) \quad (3.46)$$

Here, since we are dealing with an atom that has absorption at a certain frequency, we drop the index j that counts for different atoms and the associated summation. Thus, we can express the real and imaginary parts of $\tilde{\chi}_{abs}$ as follows:

$$(\tilde{\chi}_{abs})^r = \frac{Nq^2}{m\epsilon_0\omega_0}\left(\frac{2(\omega_0 - \omega)}{4(\omega - \omega_0)^2 + \gamma^2}\right) \quad (3.47)$$

$$(\tilde{\chi}_{abs})^i = \frac{Nq^2}{m\epsilon_0\omega_0}\left(\frac{\gamma}{4(\omega - \omega_0)^2 + \gamma^2}\right) \quad (3.48)$$

Remember the Lorentzian we discussed in Sect. 1.1.2 and note that the imaginary part of susceptibility $(\tilde{\chi}_{abs})^i$ is also a Lorentzian.

$$|F(\omega)|^2 = \frac{\beta/\pi}{\beta^2 + (\omega - \omega_0)^2} = \frac{1}{\beta\pi}\left[\frac{\beta^2}{\beta^2 + (\omega - \omega_0)^2}\right] \quad (1.38)$$

We obtained the Lorentzian (1.38) by Fourier transforming the exponentially decaying harmonic signal that was a solution to the equation of motion (1.20) for a spring–mass system of the natural frequency of ω_0. There β was the decay coefficient, and in the frequency domain, it represented the HWHM (Half-Width at Half Maximum) of the Fourier spectrum (See Table 1.1). The electric susceptibility expression $(\tilde{\chi}_{abs})^i$ results from the solution (3.26) to the equation of motion (3.23) for the electron of a dipole. So, the similarity between Eqs. (1.38) and (3.48) is not surprising.

Comparing equations of motion (1.20) and (3.23), we find $\gamma = 2\beta$. Defining FWHM (Full-Width at Half Maximum) as $\gamma \equiv \Delta\omega$, we can put $(\tilde{\chi}_{abs})^i$ in the form that explicitly indicates the spectral width.

$$(\tilde{\chi}_{abs})^i = \frac{Nq^2}{m\epsilon_0\omega_0}\left(\frac{\Delta\omega}{4(\omega - \omega_0)^2 + 4(\Delta\omega/2)^2}\right) = \frac{Nq^2}{16\pi^2 m\epsilon_0\nu_0}\left(\frac{\Delta\nu}{(\nu - \nu_0)^2 + (\Delta\nu/2)^2}\right)$$

$$(3.49)$$

Here, on the right-hand side of Eq. (3.49), we replaced the angular frequency $\omega = 2\pi\nu$ with frequency ν.

By dividing Eq. (3.47) by Eq. (3.48), we find the following relation between the real and imaginary parts of $\tilde{\chi}$:

$$\frac{(\tilde{\chi}_{abs})^r}{(\tilde{\chi}_{abs})^i} = \frac{2(\omega_0 - \omega)}{\gamma} = \frac{2(\omega_0 - \omega)}{\Delta\omega} \tag{3.50}$$

Equation (3.50) indicates that when the frequency is HWHM away from the resonant frequency the real and imaginary part of the susceptibility is equal to each other, $|\omega - \omega_0| = \Delta\omega$, $(\tilde{\chi}_{abs})^r = (\tilde{\chi}_{abs})^i$.

Figure 3.3 plots the real and imaginary parts of $\tilde{\chi}_{abs}$ as a function of frequency. As indicated by Eqs. (3.44) and (3.45), the real part is related to the index of refraction, and the imaginary part represents the absorption. The graphs are normalized to the peak value of the imaginary part at the resonance. It is seen that the real and imaginary parts are equal to each other at the frequency where the absorption becomes half of the peak value (Eq. (3.50)). Also seen is that the index of refraction has a sharp frequency dependence near the resonance frequency. This means that as far as the frequency of the incident light is outside this near resonance range the effect of dispersion is low. However, it increases fairly rapidly with the frequency from the red side of the resonance. This is good to know when designing an optical system using transmissive optics such as lenses and polarizers. We may encounter a situation where the frequency dependence of the refractive index is negligibly small for normal usage of the system but has a considerable effect when the light frequency is shifted toward the resonance. As an example, you may have a situation where a prism-type polarizer or a half-wave plate suddenly starts to perform poorly, and the problem is not the degradation of these optical components but the use of them in an improper frequency range. Do not forget that the index of refraction varies with frequency.

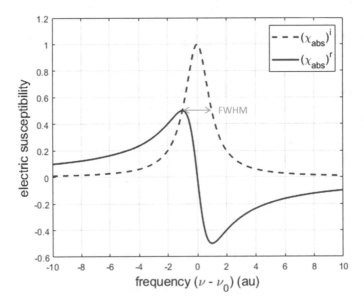

Fig. 3.3 Frequency dependence of absorption and refractive index

Another interesting feature in Fig. 3.3 is the anomalous dispersion observed on the high-frequency side of the resonance frequency. As we will discuss in Sect. 4.3, anomalous dispersion makes the group velocity higher than the phase velocity. It follows that the group velocity of light can be higher than the speed of light when it passes through a dispersive medium. While this statement is true in most media, the absorption is high in the frequency range where anomalous dispersion occurs. Therefore, the phenomenon that a light pulse travels faster than the speed of light is not easy to observe. However, this phenomenon has been observed in some materials in which the absorption in the near resonance range is low [14].

3.2.2 Amplitude of Light Wave

In Sect. 2.2.8, we discussed that the Poynting vector represents the flow of electromagnetic energy. Equations (2.81) and (2.82) indicate that the amplitude of the Poynting vector is proportional to the square of the amplitude of the electric and magnetic field, respectively. In this section, we consider the physical meaning of the amplitude of the electric or magnetic field of light waves.

Substitution of the speed of light expression $c = \sqrt{\epsilon\mu}$ into Eqs. (2.81) and (2.82) yields the following expressions of S_0:

$$S_0 = \frac{E_0^2}{c\mu} = \frac{E_0^2\sqrt{\epsilon\mu}}{\mu} = E_0^2\sqrt{\frac{\epsilon}{\mu}} \tag{3.51}$$

$$S_0 = \frac{cB_0^2}{\mu} = \frac{B_0^2}{\mu\sqrt{\epsilon\mu}} = \left(\frac{B_0}{\mu}\right)^2\sqrt{\frac{\mu}{\epsilon}} \tag{3.52}$$

Equate the right-hand sides of the above equations, rearrange the resultant equation as follows, and consider the dimension of each term:

$$E_0^2 = \left(\frac{B_0}{\mu}\right)^2\left(\sqrt{\frac{\mu}{\epsilon}}\right)^2 \tag{3.53}$$

The left-hand side is in $(V/m)^2$ as we discussed previously. From Ampère's law (Eq. (2.2), Sect. 2.1.5), $[(B/\mu)^2] = (A/m)^2$. Therefore, the dimension of the quantity $\sqrt{\mu/\epsilon}$ is $[(V/m)^2/(A/m)^2]^{1/2}$=V/A=Ω, i.e., electric resistance. This quantity is known as the intrinsic impedance of the medium z_{em}. It is the electromagnetic version of acoustic impedance or other specific impedances [15, 16].

$$z_{em} = \sqrt{\frac{\mu}{\epsilon}} \tag{3.54}$$

Knowing that $\sqrt{\mu/\epsilon}$ represents the impedance of the medium and the electric field is the voltage per unit length, we can interpret S_0 expressed in the form of Eq. (3.51) as corresponding to the power of electric current $P = V^2/Z = IV$. Here, I is the current, V is the voltage drop across the impedance Z, and Ohm's law $V = IZ$ is used. Similarly, we can interpret S_0 in the form of Eq. (3.52) as the electric power in the form of $P = IV = I^2 Z$.

Further dimensional analysis on Eq. (3.51) clarifies the physical meaning of S_0 as the optical intensity (power per unit cross-sectional area). Using $[E]=(N/C)$ (Remember $f = qE$, Eq. (2.10)) for the first E of E^2 and (V/m) for the second E, we find the unit of the right-hand side of Eq. (3.51) is $(N/C)(V/m/\Omega)$. Knowing that $V/\Omega=A=C/s$, we can express this unit as follows:

$$[S_0] = \left[E_0^2 \sqrt{\frac{\epsilon}{\mu}} \right] = \frac{(N/C)(V/m)}{\Omega} = \frac{(N/C)(C/s)}{m} = \frac{N}{(sm)} = \frac{(Nm)}{(m^2 s)} = \frac{J}{m^2 s} = \frac{W}{m^2}$$

This analysis explicitly indicates that the quantity S_0 has the unit of power per unit area, i.e., the optical power density or the intensity and that the amplitude square E_0^2 is proportional to the optical power intensity with the reciprocal of the medium constant z for the constant of proportionality.

Using Eq. (2.69), we can generally express the electric field of a light wave of angular frequency ω and wavenumber (spatial frequency) $k = 2\pi/\lambda$ traveling in the positive z-direction in the following form:

$$E(x, y, t) = E_0(x, y) \cos(\omega t \pm kz) \tag{3.55}$$

Regardless of the explicit form of the function $E_0(x, y)$, the intensity can be found at a given coordinate point (x, y) as $S_0 = E_0^2/z_{em} = E_0^2 \sqrt{(\epsilon/\mu)}$ in accordance with the above argument.

3.2.3 Phase of Light Wave

So far, we have been discussing light primarily as a carrier of electromagnetic energy. We can also view light as the propagation of a specific pattern of the electromagnetic field through space. In this regard, it is important to discuss the phase of light waves. The phase is somewhat of an abstract concept; but in the present context, we can characterize the phase of a wave as something that describes its periodically varying feature. Think of the phase of the moon. As a physical entity, the moon doesn't change its physical parameters such as its size and weight whether it is a half moon or crescent. However, its appearance varies as it revolves around the earth. Every 4 weeks or so, it appears to us with the same shape and we call it the phase of the moon. While we can see the moon once a day (except for the new moon), we need to wait for a couple of days to see the change in its appearance. If we pay attention to only full moons, the frequency becomes once a month.

The phase of a light wave can be understood as something analogous to the phase of the moon. The frequency of a wave corresponds to the frequency of a specific phase of the moon. If we associate the full moon with the phase of 2π for a cosine wave $E_0 \cos(\omega t - kz + \phi)$, as an example, the maximum (peak) value of this wave function corresponds to the full moon. Here, $\omega = 2\pi v$ is the angular frequency, k is the spatial frequency (wavenumber), and ϕ is the initial phase. We observe the maximum (peak) value E_0 every $2\pi/\omega$ s at a given place. The absolute time to observe the peak value depends on the initial phase ϕ. If $\phi = 0$, the peak value is observed at $t = 0, 2\pi/\omega, 4\pi/\omega, ...2N\pi/\omega...$ (N is an integer). If $\phi \neq 0$, the peak value is observed at different times. However, the period $2\pi/\omega$ observed at a fixed point on the z-axis is the same regardless of ϕ.

We can characterize the above wave in the space domain in the same fashion as the time domain. The spatial periodicity, called the wavelength, is defined as $2\pi/k$. By moving on the z-axis by an integer multiple of the wavelength, we observe the same phase at a given time. As is the case of the time domain argument, the phase varies with the initial phase and the absolute spatial position of observation. However, its periodicity λ remains the same regardless of ϕ or z.

The phase of a light wave contains the information called the optical path length. From a reference point, the light wave gains its phase $\theta = \omega t - kz + \phi$ as it travels along the z-axis. Therefore, if we measure the phase at two locations z_1 and z_2 at the same time, the distance is proportional to the differential optical path length $\Delta\theta \equiv \theta(z_2) - \theta(z_1) = k(z_1 - z_2)$. Since the optical wavelength is short as compared with our physical rulers, k is a large number. We can view this expression of the differential phase as a formula to magnify the physical distance $z_1 - z_2$ by a large factor k. We can apply this magnification mechanism in the technique known as optical interferometry and perform length measurements with high precision.

Optical interferometry

Interferometry [17–20] converts the phase into intensity using a reference wave of which wavelength and phase velocity are known to the user. In an interferometer, we can split the subject wave into two paths; one is called the signal wave, and the other is the reference wave. From the interference pattern, we can determine the phase. Consider a simple case where the subject wave is sinusoidal oscillating at a single frequency. After splitting it into signal and reference waves, we can express the as follows:

$$A_s = E_1 e^{i(\omega t - kl_s)} \equiv E_1 e^{i\theta_s} \tag{3.56}$$

$$A_r = E_2 e^{i(\omega t - kl_r)} \equiv E_1 e^{i\theta_r} = E_2 e^{i(\theta_s + \phi)} \tag{3.57}$$

Here, A_s and A_r are the amplitude of the signal and reference waves, and $\phi = \theta_r - \theta_s = k(l_s - l_r)$ is the phase of the reference wave relative to the signal wave. For simplicity, we set the amplitude of the signal and reference waves to be the same ($E_1 = E_2 = E_0$) but the argument here holds for $E_1 \neq E_2$. (See B.3 in Appendix.) From Eqs. (3.56) and (3.57), we find the intensity of the superposed light wave as follows:

$$I_{\text{sum}} = (A_s + A_r)(A_s + A_r)^* = E_0^2(e^{i\theta_s} + e^{i\theta_r})(e^{-i\theta_s} + e^{-i\theta_r})$$
$$= E_0^2\left(2 + e^{i(\theta_r - \theta_s)} + e^{-i(\theta_r - \theta_s)}\right) = E_0^2(2 + 2\cos(\theta_r - \theta_s))$$
$$= 2E_0^2(1 + \cos\phi) \tag{3.58}$$

Since $\cos\phi$ takes a value between -1 and 1 depending on ϕ, the total intensity varies $0 \leq I_{\text{sum}} \leq 2$ as a function of ϕ. Thus, if we somehow vary ϕ and measure I_{sum}, we can find the periodicity.

Figure 3.4 illustrates a simple interferometric optical configuration, as an example, to vary ϕ. This type of configuration is called the Michelson-type interferometer [19, 20]. The light from a source (usually a laser) is split by a beam splitter into two optical paths. The vertical optical path is for the signal light and the horizontal path is for the reference light. A reflector is placed at the end of each path so that reflected light beams are superposed when they return and merge at the beam splitter. An optical detector is placed behind the beam splitter to measure the intensity of the superposed light. The total reflector used for the reference path is moved perpendicularly to the optical path. When this reflector is moved toward the beam splitter by d, the total optical path for the reference beam is shortened by $2d$.

Since the relative phase ϕ is related to the physical length difference between the signal and reference paths Δl as $\phi = k(l_s - l_r) \equiv k\Delta l$ and $\Delta l = 2d$, we can express ϕ as follows:

$$\phi = k\Delta l = 2kd \tag{3.59}$$

Substituting Eq. (3.59) into Eq. (3.58), we can express that the intensity captured by the optical detector varies as a function of d as follows:

$$I_{\text{sum}} = 2E_0^2(1 + \cos 2kd) \tag{3.60}$$

Figure 3.5 plots the variation of I_{sum} as a function of the displacement of the reference-path end mirror d. Here, the wavelength is 500 nm and the initial ϕ (the relative phase before the total reflector is moved, i.e., $d = 0$) is set to $5/8\pi$ and the intensity is normalized to E_0^2.

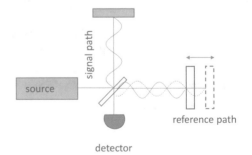

Fig. 3.4 Michelson interferometric setup to find spatial periodicity

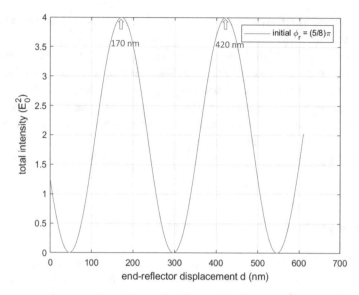

Fig. 3.5 Variation of total intensity as a function of reflector displacement

As expected, the function shows the periodicity when d varies every 250 nm, half of the wavelength.

Thus, the phase of light can be used for precise length measurement. In the above example, we used the interferometric setup to find the wavelength based on the measured reflector displacement and periodicity of the total intensity dependence on the displacement. We can use the same type of setup to find the length precisely with knowledge of the wavelength. In this case, we find the optical detector's signal I_{sum} relative to the intensity of the light source, $I_0 = E_0^2$. From Eq. (3.58), we obtain the following relation:

$$\frac{I_{sum}}{I_0} = 2(1 + \cos\phi) \tag{3.61}$$

By using the relative optical intensity measured for the left-hand side, we can solve Eq. (3.61) for ϕ. Subsequently, we can use the value of ϕ in Eq. (3.59) to solve it for l based on the known k.

Notice that the resolution of the above method of length measurement solely depends on how precisely we can evaluate Eq. (3.61). In other words, the length resolution is defined by the intensity resolution and the precision of the wavelength. Later in this Chapter (Sect. 3.1.1), we will discuss that the speed of light in a medium (including air) c is determined by the medium's index of refraction n as $c = c_0/n$ (c_0: the speed of light in vacuum). As long as the medium property is stable so that n is well defined and unchanged during the length measurement, the precision of the wavelength is determined by the frequency stability of the light source. With the use of an extremely highly stable laser source and extremely high vacuum (instead of using a stable medium), it is possible to measure length on the order of 10^{-18} m or better. [20].

We can use the phase of light for precise imaging as well. The resolution of the optical intensity has a limit known as the diffraction limit (See Eq. (4.66)). When the size of the object to be imaged is on the order of the wavelength of the light that illuminates the object, the image becomes blurry due to the phenomenon known as diffraction. On the other hand, the phase does not have a natural limit. As mentioned above, the use of high precision in the light source makes it possible to image an object based on the phase. [21].

The above arguments clearly indicate the advantage of the use of phase in precision measurement.

3.2.4   Propagation of Poynting Vector

Before ending this section, we quickly see that the Poynting vector travels at the speed of light by considering a sinusoidal wave in the form of Eq. (3.55). Using Eq. (2.75) and assuming that the E-field is polarized along the x-axis and the light is propagating in the positive z-direction, we can express the electric and corresponding magnetic-field vectors as follows:

$$\mathbf{E} = E_0(x, y) \cos(\omega t \pm kz)\hat{x} \tag{3.62}$$

$$\mathbf{B} = \frac{E_0(x, y)}{c} \cos(\omega t \pm kz)\hat{y} = \sqrt{\mu \epsilon}\, E_0(x, y) \cos(\omega t \pm kz)\hat{y} \tag{3.63}$$

Then the Poynting vector $\mathbf{S} = \mathbf{E} \times \mathbf{B}/\mu$ becomes

$$S = \frac{E_0^2(x, y)\sqrt{\epsilon}}{\sqrt{\mu}} \cos^2(\omega t - kz) = \frac{E_0^2(x, y)}{2} \frac{1}{z_{em}} \big[1 + \cos(2\omega t - 2kz)\big] \tag{3.64}$$

Alternatively, using $E_0 = cB_0 = B_0/\sqrt{\epsilon \mu}$ to eliminate E_0

$$S = \frac{B_0^2(x, y)}{\mu\sqrt{\epsilon \mu}} \cos^2(\omega t - kz) = \frac{z_{em}}{2} \left(\frac{B_0(x, y)}{2\mu}\right)^2 \big[1 + \cos(2\omega t - 2kz)\big] \tag{3.65}$$

Here, z_{em} is the specific impedance (3.54). Note that the cosine term on the rightmost-hand side of Eqs. (3.64) and (3.65) indicates that the phase velocity of $\mathbf{S}$ is still $c = 1/\sqrt{\epsilon \mu}$ as $2\omega/2k = c$. Equation (3.64) represents the intensity flow as "voltage (force)-like" wave and Eq. (3.65) as "current (velocity)-like" wave.

3.2.5 Polarization of Optical Waves

Polarized light is light whose electric field vector is restricted to a certain plane [22]. The plane waves we discussed above are all linearly polarized because the electric field oscillates along a specific line as the wave propagates. Figure 3.6 illustrates the situation where the electric field of a linearly polarized light wave is expressed with an xyz-coordinate system. It is a snapshot of the electric field of the light. The linearly polarized light propagates in the positive z-direction. The sine-shaped line on the zx plane represents the x-component of the electric field, $E_x\hat{i}$. Similarly, the one on the yz plane represents the y-component, $E_y\hat{j}$. The third sine-shaped line is the total electric field, $\mathbf{E} = E_x\hat{i} + E_y\hat{j}$. Since the x- and y-components have the same amplitude, $E_x = E_y$, the total electric field makes an angle of $45°$ to the x- and y-axes.

The solid straight lines connecting the sine-shaped line of the total electric field and the z-axis depict the direction of the electric field vector $\mathbf{E} = E_x\hat{i} + E_y\hat{j}$ at every point along the axis of propagation with an increment of $\pi/4$ in phase. All these lines are in the same plane, indicating that the electric field of the linearly polarized light remains restricted to the same plane of propagation. Each of these straight lines represents the total electric field of the light that leaves the light source at the time corresponding to a phase of $\pi/4$ earlier than the neighboring straight line on the left. For example, the second straight line from $z = 0$ leaves $\Delta t = (\pi/4)/\omega$ earlier than the first straight line. This observation indicates that the electric field is always at $45°$ to the x and y axes as the light travels.

A linearly polarized light remains in the same plane as shown in Fig. 3.6 under two conditions. The first condition is that the light source emits a linearly polarized wave. Natural light such as one from a light bulb is unpolarized, or randomly polarized. If we draw a picture equivalent to Fig. 3.6 for a randomly polarized light source, the lines representing the field vectors would not remain in the same plane.

The second condition is that the index of refraction is uniform in the medium which the linearly polarized light propagates through. As discussed in the preceding section, the phase of light advances as kz, where k is the propagation constant and z is the coordinate variable for the axis of propagation. The propagation constant in a medium is given as $k = k_0 n$ where k_0 is the propagation constant in a vacuum and n is the index of refraction. If the index of

Fig. 3.6 Linearly polarized light expressed with a coordinate system whose x-axis and y-axis have $45°$ to linear polarization

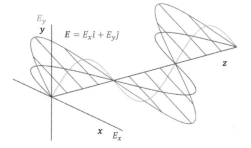

refraction is uniform in the transmissive medium, the phase advancement depends only on the thickness of the medium. It does not depend on the orientation of the field vector; hence, the vector components advance the phase equally. Consequently, the orientation of the vector is unchanged, i.e., the transmitted light has the same linear polarization as the incident light.

On the other hand, if the index of refraction along the x-axis is different from the y-axis inside the transmissive medium, the total phase that the wave advances for the same physical distance is different between the x-component, E_x, and the y-component, E_y. The property that the medium has two different indexes of refraction along two axes is referred to as the birefringence [23, 24], and a medium that has this property is called a birefringent medium. In a birefringent medium, the axis along which the index of refraction is lower, and consequently the light propagates faster, is called the fast axis [26], and the other axis is called the slow axis [26]. If the x-axis is the fast axis, $n_x < n_y$. The fast axis is also referred to as the ordinary axis and the slow axis as the extraordinary axis, i.e., the index of refraction is lower for the ordinary axis. $n_o < n_e$.

A birefringent medium can turn linear polarization into circular or elliptical polarization. Consider how the field vector components behave when they pass through a birefringent medium. Figure 3.7 illustrates the situation where a linearly polarized light incident to a birefringent disk becomes a circularly polarized light after being transmitted through the disk. Let the x-axis be the fast axis and the y-axis be the slow axis. Prior to the entrance to the birefringent disk, the x- and y-components are in phase. When one is at its peak, so is the other. After the transmission, the two components become out of phase. The x-component starts off with its peak, whereas the y-component starts off with zero. The birefringent disk creates a phase difference of $\pi/2$, which makes the x-component of the transmitted light behave cosine-like and the y-component behave negative sine-like. This phase difference of $\pi/2$ remains the same as the wave propagates because the index of refraction becomes uniform again after the transmission.

Figure 3.8 depicts the mechanism with which the birefringent disk produces the phase difference of $\pi/2$. Throughout the thickness of the disk, the slow-axis component (the vertical component) undergoes three full oscillations, whereas the fast-axis component undergoes

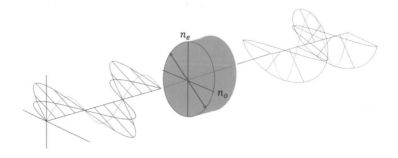

Fig. 3.7 Linear polarization turns into circular polarization as light passes through birefringent disk

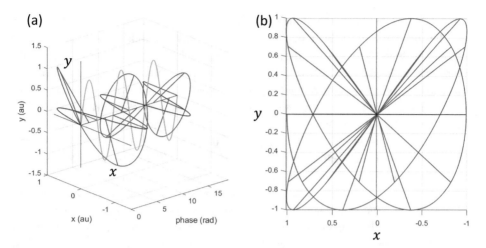

Fig. 3.8 Phase difference between fast and slow axes inside the birefringent disk. The x- and y-components of the electric field are initially in phase. The phase of the slow-axis component is advanced by $\pi/2$ at the end of the disk

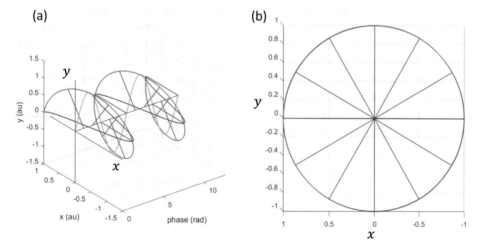

Fig. 3.9 Circularly polarized light output from a quarter plate. The x- and y-components of the electric field keep the $\pi/2$ phase difference produced by the quarter-wave plate

two and a quarter oscillations. In Fig. 3.8a, the straight line drawn every $\pi/4$ in the slow axis phase depicts the total field vector. Unlike Fig. 3.6, these straight lines do not form a plane. Instead, it rotates around the axis of propagation. Figure 3.8b illustrates the change of the total vector in its orientation and length viewed from behind the light wave. Notice that the phase advances faster vertically than horizontally. The length of the vector is determined by the resultant orientation of the vector at each phase.

Figure 3.9 is an illustration of the transmitted light's vector in the same format as Fig. 3.8. This time the vertical and horizontal phases advance at the same rate and therefore the length of the total vector remains the same. The phase advancement of $\pi/2$ for the vertical component makes the trajectory of the tip of the vector circular. This type of polarization is referred to as circular polarization [22]. A birefringent component that produces a phase difference of $(2N + 1/2)\pi$, where N is an integer, is called a quarter-wave plate [25]. The word "quarter" is used because $\pi/2$ is a quarter of the full period 2π.

When a birefringent disk produces a phase difference of π between the fast and slow axes, the disk is called a half-wave plate [25]. Figure 3.10 illustrates the behaviors of the fast-axis and slow-axis components of a linearly polarized light incident to a half-wave plate. Here, Fig. 3.10a shows how the fast and slow-axis components behave inside the half-wave plate. While the slow-axis (vertical) component makes four full oscillations, the fast-axis (horizontal) component makes three and a half oscillations. This breaks the linear polarization inside the plate and causes the net phase difference between the axes to be π at the exit surface of the plate. This relative phase change flips the sign of the fast-axis component at the exit surface of the plate as compared with at the entrance surface as seen at the far end of Fig. 3.10a.

The above situation makes the polarization of the transmitted light linear again. Figure 3.10b shows the trajectory of the tip of the field vector in the same format as Fig. 3.9b. Here, the solid line is for the transmitted light and the dashed line is for the incident light. Because the horizontal component's amplitude is flipped at the exit surface, the orientation of the linear polarization of the transmitted light is symmetric to that of the incident light about the vertical axis. In Fig. 3.10b, θ_{in} and θ_{tr} are, respectively, the angle of the incident and transmitted light's polarization to the y-axis. The symmetry about the slow (y)-axis is represented by $\theta_{in} = \theta_{tr}$.

The above-mentioned symmetry about the slow axis indicates that the angle between the planes of the incident light's polarization and the transmitted light's polarization is equal to twice the angle made by the incident light's polarization and the y-axis. In other words, the half-wave plate rotates the polarization of the incident light by angle $\theta_r = 2\theta_{in}$.

$$\theta_r = \theta_{in} + \theta_{tr} = 2\theta_{in} \qquad (3.66)$$

The effect represented by Eq. (3.66) is widely used to rotate the orientation of linear polarization. Notice that the angle of rotation θ_r can be controlled by the angle of the incident polarization to the slow axis. Naturally, the same argument holds for the angle of the incident polarization with respect to the fast axis. The angle of rotation is twice the angle of the fast axis.

If the phase difference between the fast- and slow-axis components deviates from π by extra phase δ as $\pi + \delta$, the transmitted light's polarization becomes elliptical. Figure 3.10c shows the fast- and slow-axis components inside the disk when the extra phase is 10%, i.e., the net phase difference is equal to $\pi + \delta = 1.1\pi$. As you can easily imagine, if the extra phase δ is increased from 10% the ellipse bulges. When $\pi + \delta$ becomes $\pm\pi/2$ the elliptical

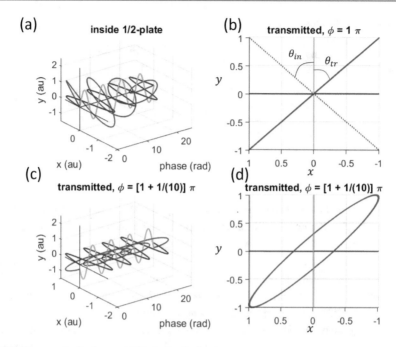

Fig. 3.10 Linear polarization and elliptical polarization

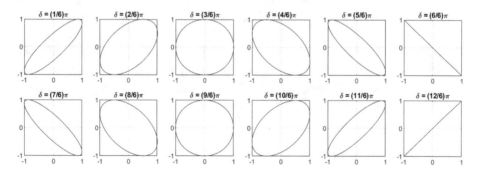

Fig. 3.11 Transmitted light's polarization vs extra phase δ

polarization becomes a circular polarization. If δ is further increased, the transmitted light's polarization starts to contract. At the same time, the major axis of the ellipse starts rotating in the opposite direction. Figure 3.11 illustrates how the polarization varies with δ.

There are many occasions where defining the polarization well is significant. Here, we consider two cases. The first is the management of light waves at the boundary of media having different indexes of refraction. As we elaborate in Sect. 4.1, the reflection and transmission characteristics of light depend on the polarization. Conventionally, we define the polarization at a boundary in association with the plane of incidence. Here, the plane of

Fig. 3.12 Plane of incidence.
In this example, the electric
field is *s*-polarized and the
magnetic field is *p*-polarized

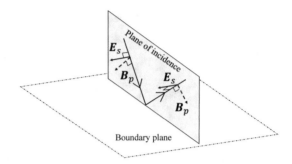

incidence is defined as the plane that contains the propagation vector of the incident light
and reflected light. Figure 3.12 illustrates the plane of incidence along with the *p*- and *s*-
polarization schematically. When the electric-field vector is parallel to the plane of incidence,
the polarization is referred to as *p*-polarized [27]; if the electric-field vector is perpendicular
to the plane of incidence, the polarization is referred to as *s*-polarized [27]. Here, "s" of *s*-
polarization comes from the German word *senkrecht*, which means perpendicular. Similarly,
"p" comes from the German word *parallel*, whose meaning is obvious.

The second case where polarization is important is interference. In engineering, a number
of interferometric techniques are used for various purposes [18]. As discussed above under
"Phase of light wave", optical interferometry evaluates the phase of light using the intensity
of the total optical field resulting from the superposition of the interfering light (component)
waves. Being a vector field, the optical field interferes with each other via vector additions
of the component waves. In other words, the superposition takes place only for the same
polarization. A *p*-polarized component wave and an *s*-polarized component wave do not
interfere with each other. Consequently, the phase cannot be evaluated. Proper adjustment of
component waves' polarization is essential to increase the accuracy of the phase evaluation.

3.3   Light Propagation

3.3.1   Paraxial Wave Approximation and Helmholz Equation

A paraxial wave is a wave whose intensity is localized in the vicinity of the axis of propagation.[3] What does the word "vicinity" mean? It is defined by the source of the wave.
When a wave is generated with a source, the source has a finite dimension transverse to the
axis of the wave propagation. It is impossible to generate a wave outside of this transverse
dimension of the source. This defines the starting (the initial) transverse size of the wave.
The vicinity is naturally defined by this initial transverse size of the wave.

---

[3] See [28] for solving the Helmholtz equation with Paraxial approximation.

When the wave source has a finite transverse size, the wave exhibits natural diffraction. This causes two main differences from the plane wave case. First, the radius of the wave (beam) increases as the wave propagates, Second, the wavefront is curved. The former is obvious from the definition of diffraction. The latter may not be as obvious but is intuitively understood by considering a little rock thrown into a calm pond. This small wave source will generate a circular wavefront that spreads concentrically.

Mathematically, a paraxial solution is given as a solution to a differential equation known as the Helmholtz equation [29]. As will be clarified below, with certain approximations, the solution exhibits a transverse profile of the amplitude in the form of Gaussian distribution. Thus, the solution is generally called a Gaussian beam. Since the approximations are used in the solution process, the resultant Gaussian beam is a paraxial approximation of the wave. In most applications, this approximation is accurate enough to describe the wave propagation and related phenomena such as the focusability of the wave and the focal size. A Gaussian beam solution tells us the beam size (the wave radius) and the wavefront (the radius of curvature) at a given point on the propagation axis.

Let's start with a three-dimensional wave equation in the form of (1.89).

$$\frac{\partial^2 \xi}{\partial t^2} = \frac{K}{\rho} \nabla^2 \xi \tag{3.67}$$

Here, $K = \lambda + 2G$ is the elastic modulus. First, separate the solution function $\xi(t, x, y, z)$ into a time-dependent and space-independent function $T(t)$, and a time-independent and space-dependent function $S(x, y, z)$.

$$\xi(t, x, y, z) = \xi_0 T(t) S(r) \tag{3.68}$$

Here, r is the radial distance of a coordinate point from the origin; $r = \sqrt{x^2 + y^2 + z^2}$. Substitution of (3.68) into differential equation (1.89) and rearrangement of terms yield the following equation:

$$\frac{\ddot{T}}{T} = v_p^2 \frac{\nabla^2 S}{S} = -\omega^2 \tag{3.69}$$

Here, K/ρ in Eq. (1.89) has been replaced with the square of the phase velocity v_p defined by Eq. (1.90). The rightmost-hand side of Eq. (3.69) comes from the assumption that the oscillation is in the form of

$$T(t) = T(0)e^{i\omega t} \tag{3.70}$$

Expressing v_p as the ratio of the temporal to spatial frequency (See Eq. (1.87)), we can rewrite Eq. (3.69) in the following form, which is known as the Helmholtz equation:

$$\nabla^2 S + k^2 S = 0 \tag{3.71}$$

Here, we can identify k as the magnitude of the propagation vector defined by Eq. (1.78) assuming that $S(r)$ has a spatially sinusoidal variation. For simplicity, set the z-axis in the direction of the propagation. In this case, we can put $S(r)$ in the following form:

$$S(r, z) = \Psi(x, y, z)e^{-ikz} = \Psi(r, z)e^{-ikz} \tag{3.72}$$

Here we assume Ψ is cylindrically symmetric about the z-axis. The expression $\Psi(r, z)$ emphasizes this spatial dependence of Ψ.

Substitute expression (3.72) into the differential equation (3.71). In doing so, first split the Laplacian into the radial and axial components as follows:

$$\nabla^2 = \frac{\partial^2}{\partial r^2} + \frac{1}{r}\frac{\partial}{\partial r} + \frac{\partial^2}{\partial z^2} \tag{3.73}$$

Using Eq. (3.73), we can rewrite the Laplacian part of the differential equation (3.71) in the following form:

$$
\begin{aligned}
\nabla^2 S &= \left(\frac{\partial^2 \Psi}{\partial r^2} + \frac{1}{r}\frac{\partial \Psi}{\partial r}\right)e^{-ikz} + \frac{\partial^2(\Psi e^{-ikz})}{\partial z^2} \\
&= \left(\frac{\partial^2 \Psi}{\partial r^2} + \frac{1}{r}\frac{\partial \Psi}{\partial r} - 2ik\frac{\partial \Psi}{\partial z} - k^2\Psi + \frac{\partial^2 \Psi}{\partial z^2}\right)e^{-ikz} \\
&\cong \left(\frac{\partial^2 \Psi}{\partial r^2} + \frac{1}{r}\frac{\partial \Psi}{\partial r} - 2ik\frac{\partial \Psi}{\partial z} - k^2\Psi\right)e^{-ikz}
\end{aligned} \tag{3.74}
$$

Here, in the last step, we assume that the function Ψ varies along the z-axis slowly, and neglect the secondary differentiation of Ψ with respect to z.

Substitution of expression (3.74) into the differential equation (3.71) results in the following equation:

$$\nabla^2 S + k^2 S = \left(\frac{\partial^2 \Psi}{\partial r^2} + \frac{1}{r}\frac{\partial \Psi}{\partial r} - 2ik\frac{\partial \Psi}{\partial z}\right)e^{-ikz} = 0 \tag{3.75}$$

Since Eq. (3.75) holds for any z, we obtain the following differential equation:

$$\frac{\partial^2 \Psi}{\partial r^2} + \frac{1}{r}\frac{\partial \Psi}{\partial r} - 2ik\frac{\partial \Psi}{\partial z} = 0 \tag{3.76}$$

By solving Eq. (3.76), we can find an expression for the spatial part of the electric field of a light propagating in the positive z-direction.

To solve differential equation (3.76), we set the function $\Psi(r, z)$ in the following form:

$$\Psi(r, z) = \Psi_0 e^{-i\left[P(z) + \frac{1}{2}Q(z)r^2\right]} \tag{3.77}$$

With the mathematical operations described in Appendix C.2, we obtain the following expressions for functions $P(z)$ and $Q(z)$:

$$P(z) = -i \ln(1 + \frac{z}{q_0}) \tag{3.78}$$

$$Q(z) = \frac{k}{z + q_0} \tag{3.79}$$

With further mathematical operations, which are detailed in Appendix C.2, we can eventually express the space part of the paraxial wave in the following form:

$$\Psi(r, z) = \Psi_0 \frac{w_0}{w_0\sqrt{1 + \left(\frac{z}{z_0}\right)^2}} e^{i \tan^{-1}\left(\frac{z}{z_0}\right)} e^{-r^2\left(\frac{1}{w(z)} + \frac{ik}{2R(z)}\right)} = \Psi_0 \frac{w_0}{w(z)} e^{-\frac{r^2}{w(z)^2}} e^{i\phi_G} e^{-i\frac{kr^2}{2R(z)}}$$

$$\tag{3.80}$$

The first exponential term on the rightmost-hand side of Eq. (3.80) takes the form of a Gauss' error function, indicating that the radial dependence of the amplitude of $\Psi(r, z)$ is a Gaussian distribution. That is why, we call this paraxial wave the Gaussian wave. The second exponential term indicates that the on-axis phase of Ψ is equal to the parameter ϕ_G. It is a constant at each z location, determined by z_0 (see below). The third exponential term indicates that Ψ's radial phase varies quadratically depending on $R(z)$. Since $R(z)$ depends only on z, we find that Ψ's wavefront is a sphere whose radius of curvature is R. These observations indicate that the Gaussian beam Ψ is a spherical wave, whose amplitude has a Gaussian distribution. The width of the Gaussian distribution $w(z)$ and the wavefront's radius of curvature $R(z)$ vary as a function of axial distance z.

The quantities introduced in the above expressions are defined as follows. Appendix C.2 describes how we introduce these quantities in this formalism. Since these are important parameters to characterize a Gaussian beam, we discuss them here.

$$z_0 = \frac{\pi w_0^2}{\lambda} \tag{3.81}$$

$$q_0 = i z_0 = i \frac{\pi w_0^2}{\lambda} \tag{3.82}$$

$$w(z) = w_0\sqrt{1 + \left(\frac{z}{z_0}\right)^2} = w_0 \sqrt{1 + \left(\frac{\lambda z}{\pi w_0^2}\right)^2} \tag{3.83}$$

$$R(z) = z\left[1 + \left(\frac{z_0}{z}\right)^2\right] = z\left[1 + \left(\frac{\pi w_0^2}{\lambda z}\right)^2\right] \tag{3.84}$$

$$\phi_G = \tan^{-1}(z/z_0) \tag{3.85}$$

Here, λ is the wavelength, z_0 is the Rayleigh length, $w(z)$ is the spot size, $R(z)$ is the radius of curvature of the wavefront, and ϕ_G is the Gouy phase.

We can view q_0 as the value of the complex radius $q(z)$ at $z = 0$, $q_0 = q(0)$. The complex radius is defined based on function $Q(z)$ expressed by Eq. (3.79) as follows:

$$q(z) = \frac{k}{Q(z)} = q_0 + z \qquad (3.86)$$

Here, $k = 2\pi/\lambda$ is the wave number.

The physical significance of these quantities is as follows. From Eq. (3.80), we find that the spot size is the radial size where the amplitude becomes $1/e$ of the peak value at $r = 0$. Equation (3.83) tells us that the Gaussian beam takes the minimum spot size w_0 at $z = 0$. This minimum spot size is called the beam waist size. Equation (3.83) also tells us that the spot size is symmetric along the z-axis about $z = 0$. In other words, the Gaussian beam has the beam waist at $z = 0$, and its spot size increases as the wave travel either in the positive or negative z-direction. From Eqs. (3.83) and (3.84), we can characterize the Rayleigh length z_0 as the on-axis distance at which the spot size becomes a factor $\sqrt{2}$ greater than the beam waist and the radius of curvature is at the minimum. Equation (3.85) indicates that the Gouy phase varies on the z axis at a rate determined by the Rayleigh length. From Eq. (3.80), we can interpret the Gouy phase as the on-axis phase in excess of the phase of the plane wave of the same wavelength.

Figure 3.13a illustrates $w(z)$ as a function of radial distance from the beam axis. As Eq. (3.80) indicates, $\Psi(r, z)$ represents a Gaussian distribution of the wave's amplitude at a given axial position z. For $z = 0$ m, the spot size is indicated. Figure 3.13b illustrates that the spot size $w(z)$ increases in going either direction along the z-axis from $z = 0$ (Eq. (D.20)). That is why $w(0) = w_0$ is referred to as the beam waist. The spot size indicated in Fig. 3.13a is in fact the beam waist size as this Gaussian beam has the beam waist at the axial position $z = 0$ m. Figure 3.13c shows that the radius of curvature $R(z)$ takes the minimum value at the Rayleigh length and keeps increasing from there. The asymptotic behavior of $R(z)$ will be discussed shortly.

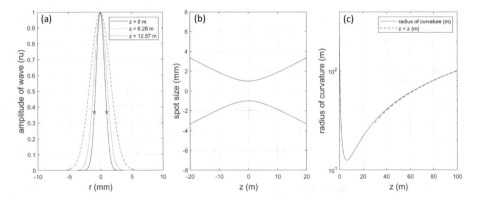

Fig. 3.13 Gaussian beam **a** Gaussian amplitude profile at three axial locations (z). Spot size (beam waist size) is indicated for $z = 0$ m. **b** Beam spot size variation along beam axis; **c** Radius of curvature variation

Adding the $exp(-ikz)$ term, we can express the entire space part of the solution (3.72) as follows:

$$S(r, z) = \Psi(r, z)e^{-ikz} = \Psi_0 \frac{w_0}{w(z)} e^{-\frac{r^2}{w(z)^2}} e^{i\tan^{-1}\left(\frac{z}{z_0}\right)} e^{-i\frac{kr^2}{2R(z)}} e^{-ikz}$$

$$= \Psi_0 \frac{w_0}{w(z)} e^{\left[-i(kz - \phi_G) - r^2\left(\frac{1}{w(z)} + \frac{ik}{2R(z)}\right)\right]} \tag{3.87}$$

Expression (3.87) along with the above discussions tells us that once we know the wavelength and the size and location of the beam waist, we can fully characterize the Gaussian beam.

3.3.2 Gaussian Optical Beam

In the preceding section, we introduced the Gaussian wave as a solution to the Helmholtz equation with the paraxial approximation. In this section, we discuss Gaussian optical beams for practical applications. Laser sources we use in our experiments and other practical purposes usually output a Gaussian beam expressed by Eq. (3.87) [30, 31]. In the context of laser applications, we often call this form of Gaussian beam the fundamental mode [32]. Under some conditions, lasers can generate a Gaussian beam of higher order modes [33–36]. We will start the discussions by connecting the concepts discussed in the preceding section for a Gaussian beam of the fundamental mode with experimentally observable quantities. The discussions in this part include practical ways to find the spot size and radius of curvature at a given distance from the laser source when the laser beam propagates through free space and a space with transmissive optics. Then we will discuss briefly higher order Gaussian modes.

Applying Eq. (3.87) to the electromagnetic field, we obtain the following expression:

$$E(r, z) = E_0 \frac{w_0}{w(z)} e^{-i\left(kz - \tan^{-1}\left(\frac{z}{z_0}\right)\right)} e^{-r^2\left(\frac{1}{w(z)^2} + \frac{ik}{2R(z)}\right)} \tag{3.88}$$

Here, $w(z)$ and $R(z)$ represent the spot size and wavefront radius of curvature of the amplitude of the electric field. Equation (3.88) indicates that we can find the radial variation of the amplitude and phase of the electric field at a given on-axis location z.

In practice, our eyes and optical power meters are sensitive to the intensity, not the amplitude of the optical field. As discussed in Sect. 3.2.2, the optical intensity is proportional to the square of the amplitude. This indicates that the spot size $w(z)$ in intensity represents the radial distance where the intensity drops to $(1/e)^2 \cong 14\%$ of the on-axis peak value. In other words, the trace of the spot size shown in Fig. 3.13b represents the radial distance where the intensity drops to 14% of the peak value. This radial distance approximately represents the beam radius when we shoot a laser beam into a wall. Thus, moving a reflective planar material along the beam axis and measuring the beam radius at various locations would allow us to estimate the beam waist size and location. (For visible lasers, a white business card usually works well for this purpose.) Note that once we find the beam waist size and location, we can find the radius of curvature at a given on-axis location.

For laboratory uses and engineering applications of a laser source, we need a more accurate estimation of the spot size and radius of curvature. Here, we discuss a convenient way for such an estimation [37].

From Eqs. (3.83) and (3.84), we find the following equality:

$$\frac{1}{w(z)^2} = \frac{1}{w_0^2\left((1+(z/z_0)^2)\right)} = \frac{z_0^2}{w_0^2(z_0^2+z^2)} = \frac{\pi w_0^2}{\lambda}\frac{z_0}{w_0^2(z_0^2+z^2)} \tag{3.89}$$

$$\frac{1}{R(z)} = \frac{1}{\left(z(1+(z_0/z)^2)\right)} = \frac{z}{z^2+z_0^2} \tag{3.90}$$

Using Eqs. (3.89) and (3.90), we find

$$\frac{1}{R(z)} - \frac{i\lambda}{\pi w(z)^2} = \frac{z}{z^2+z_0^2} - i\frac{z_0}{z^2+z_0^2} = \frac{1}{z+iz_0} \tag{3.91}$$

From Eqs. (3.82) and (3.86), we obtain the following expression for $q(z)$:

$$q(z) = q_0 + z = iz_0 + z \tag{3.92}$$

The quantity $q(z)$ is called the complex radius of a Gaussian wave.

Applying Eq. (3.92) to the on-axis location $z + d$, we find the following:

$$q(z + d) = iz_0 + (z + d) = (iz_0 + z) + d = q(z) + d \tag{3.93}$$

Equation (3.93) indicates that as we move on the beam axis by distance d, q increases by d.

Using Eqs. (3.91) and (3.92), we can express the complex radius at z with the spot size $w(z)$ and wavefront radius of curvature $R(z)$.

$$\frac{1}{q(z)} = \frac{1}{R(z)} - i\frac{\lambda}{\pi w(z)^2} \tag{3.94}$$

Equation (3.94) is a convenient expression to find the beam spot size and radius of curvature of a Gaussian wave at a given z. From Eq. (3.92), we can find the complex radius at a given z location based on the Rayleigh length z_0. Once we know $q(z)$, we can find $w(z)$ and $R(z)$.

While expressions (3.83) and (3.84) are useful to find the beam radius and radius of curvature in free space, they are not convenient when a Gaussian beam passes through optical components. This is because the z dependences of the spot size and wavefront curvature are altered on each optical component. In such cases, the use of Eq. (3.94) in conjunction with the ray (ABCD) matrix [38] is very useful. In the next section, we will discuss how to use the proper ray matrix for typical optical components. We will apply the concept to a simple case and evaluate the beam spot size and wavefront curvature as a function of z.

3.3.3 Ray Matrix

When we use a laser beam in experiments, it is likely that the laser beam passes through various optical components, such as lenses and transmissive optics. In such cases, the use of complex radius q is very powerful [37]. In most cases, the laser beam is in the fundamental Gaussian mode [32] and the optical components do not convert the mode into higher Gaussian modes. Instead, they change the propagation characteristics of the Gaussian beam. As mentioned above, a Gaussian beam is characterized by a pair of $w(z)$ and $R(z)$. Equation (3.94) allows us to express the $w(z)$-$R(z)$ pair at a given z as a single parameter $q(z)$. (As will be discussed shortly, the Helmholtz equation with the paraxial approximation can yield transverse higher order modes. The modes are indicated by the index m and n, where m represents the mode in the x and n in the y. The fundamental mode is the case when $m = n = 0$, and often called the TEM$_{00}$ mode. Here, TEM stands for Transverse Electromagnetic. Hereafter, we use TEM$_{00}$ for the fundamental Gaussian mode.)

Thus, by using $q(z)$ properly, we can express the propagation of a Gaussian beam comprehensively. The use of ray matrix allows us to evaluate the output complex radius q_{out} based on the incident value q_{in} for each optical element including the free space between optical components.

For a pair of q_{in} and q_{out}, the ray matrix takes the following form:

$$q_{out} = \frac{Aq_{in} + B}{Cq_{in} + D} \tag{3.95}$$

where each optical component has a unique expression for A, B, C, and D. Table 3.1 lists the ray matrix of typical optical components.

First, apply the ray matrix of free space. Let q_i be the initial complex radius and q_f be the final complex radius after traveling in free space a distance d. According to Table 3.1,

$$q_f = \frac{Aq_i + B}{Cq_i + D} = \frac{q_i + d}{0 + 1} = q_i + d \tag{3.96}$$

Notice that Eq. (3.96) is equivalent to Eq. (3.93).

As another example, consider the complex radius after a TEM$_{00}$ beam propagating through a positive lens of focal length f. From Table 3.1, the ray matrix of a positive (thin) lens is as follows:

$$\begin{pmatrix} A & B \\ C & D \end{pmatrix} = \begin{pmatrix} 1 & 0 \\ -\frac{1}{f} & 1 \end{pmatrix} \tag{3.97}$$

Thus, in this case, Eq. (3.95) becomes

Table 3.1 Ray matrix for typical optical components

Component	$\begin{pmatrix} A & B \\ C & D \end{pmatrix}$	Beam propagation
Free space	$\begin{pmatrix} 1 & d \\ 0 & 1 \end{pmatrix}$	
Thin lens	$\begin{pmatrix} 1 & 0 \\ \frac{-1}{f} & 1 \end{pmatrix}$	
Boundary n_1 to n_2	$\begin{pmatrix} 1 & 0 \\ 0 & \frac{n_1}{n_2} \end{pmatrix}$	

$$q_{out} = \frac{q_{in}}{-\frac{1}{f}q_{in} + 1} \tag{3.98}$$

Hence,

$$\frac{1}{q_{out}} = \frac{-\frac{1}{f}q_{in} + 1}{q_{in}} = -\frac{1}{f} + \frac{1}{q_{in}} \tag{3.99}$$

Equation (3.99) indicates that a thin lens changes the real part of $1/q(z)$ by $-1/f$ but does not affect the imaginary part of $1/q(z)$. From Eq. (3.94), we can find the beam size and wavefront radius immediately after the beam passes through the lens.

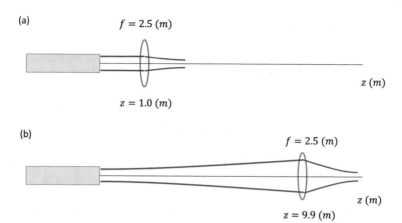

Fig. 3.14 He–Ne laser focused by f = 2.5 m lens

$$\frac{1}{R_{out}} = \frac{1}{R_{in}} - \frac{1}{f} \tag{3.100}$$

$$w_{out} = w_{in} \tag{3.101}$$

Since $f > 0$, Eq. (3.100) indicates that $1/R_{out} < 1/R_{in}$, i.e., $R_{out} > R_{in}$. The sign convention of the radius of curvature is as follows. (See Sect. 2.4 of [28].) If the center of the curvature is on the negative side of the point where the wavefront crosses the z-axis (i.e., the wavefront bulges toward the positive side of z), $R > 0$. Therefore, if a diverging beam passes through the positive lens from the negative side of the z-axis, the $R_{out} > R_{in}$ means that the initially positive R increases positively, i.e., the wavefront becomes flatter. If the incidence beam is flat, $1/R_{out} = 1/\infty - 1/f < 0$; hence, the output beam is converging. Equation (3.101) indicates that the beam size does not change, indicating that the lens does not change the beam size.

As an example, consider a TEM_{00} beam from a helium–neon laser (wavelength $\lambda =$ 632.8 nm) focused by a positive lens of focal length 2.5 m. Figure 3.14 illustrates two configurations. In configuration (a), the positive lens is placed 1 m away from the beam waist of the laser. In configuration (b), the same lens is placed 9.9 m away from the beam waist of the same laser as (a). Here, 9.9 m is twice the Rayleigh length of the TEM_{00} output from the laser.

Figure 3.15 shows the beam profile resulting from configuration (a) in Fig. 3.14. Here, Fig. 3.15a illustrates the beam spot size as a function of z whose origin is at the laser's beam waist location. The dashed line represents the beam spot size of the laser and the solid line represents the spot size after the beam passes through the positive lens. The focal position of the beam is found to be $z = 3.15$ m, and the minimum spot size there is 0.48 mm, as shown in Fig. 3.15b. At the entrance to the lens, the laser beam has a spot size of 1.02 mm and a radius of curvature of 25. 6 m (Fig. 3.15c). Figure 3.15 (d) is the radius of curvature near the focal position. It indicates that the radius of curvature is at the maximum at the focal position.

Figure 3.16 shows the same type of information as Fig. 3.15 for configuration (b) shown in Fig. 3.14. In this case, the laser beam enters the positive lens with a spot size of 2.24 mm and a radius of curvature of 12.4 m (Fig. 3.16a and c). The minimum spot size at the focal position is 0.28 mm, which is approximately half of configuration (a). Notice that, while the minimum beam size at the focal point is smaller, the spot size increases faster than configuration (a).

Table 3.2 summarizes the above-discussed He–Ne laser's focusing behavior as the lens location is varied. Here, in addition to the two configurations indicated in Fig. 3.14, the case when the positive lens is placed very close to the beam waist ($z = 0.02$ m) is included. The beam divergence angle indicates the diverging property after the focus due to the positive lens. Notice that according to Eq. (4.68), the divergence angle increases in proportion to the reciprocal of the minimum spot size at the focus.

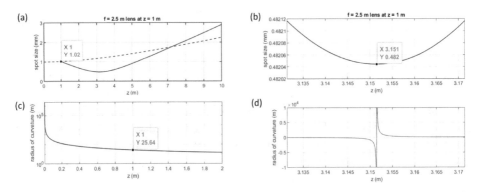

Fig. 3.15 He–Ne laser focused by f = 2.5 m lens placed at 1 m away from beam waist. **a** Beam spot size from the initial waist. **b** Focal position and spot size at focus. **c** Variation of spot size after lens. **d** Radius of curvature near focus

Table 3.2 Beam size and propagation properties of He-Ne laser focused at different positions by same lens

Lens location (m)	0.02	1.0	9.9
Spot size at lens (mm)	1.0	1.02	2.24
Spot size at focus (mm)	0.45	0.48	0.28
Beam divergence angle (rad)	4.48×10^{-4}	4.20×10^{-4}	7.19×10^{-4}

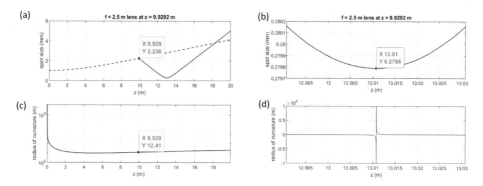

Fig. 3.16 He–Ne laser focused by f = 2.5 m lens placed at 9.9 m (twice of Rayleigh length) away from beam waist. **a** Beam spot size from the initial waist. **b** Focal position and spot size at focus. **c** Variation of spot size after lens. **d** Radius of curvature near focus

3.3.4   Higher Order Hermit-Gaussian Modes

In Sect. 3.3.1, we solved the Helmholtz equation with the paraxial approximation assuming the cylindrical symmetry around the axis ($\partial/\partial\phi = 0$, where ϕ is the azimuthal angle of the cylindrical coordinate system). If we remove this cylindrical symmetry, the Helmholtz equation (3.71) yields the following solutions known as Hermit–Gaussian modes [36, 39].

$$E_{m,n} = E_0 \frac{w_0}{w(z)} H_m\left(\frac{\sqrt{2}x}{w(z)}\right) H_n\left(\frac{\sqrt{2}y}{w(z)}\right) exp\left[-\frac{x^2 + y^2}{w^2(z)}\right]$$
$$exp\left[-\frac{ik(x^2 + y^2)}{2R(z)} - ikz + i(m + n + 1)\phi_G\right] \quad (3.102)$$

Here, H_m and H_n are the mth and nth-order Hermite polynomials [40], $w(z)$ is the spot size at z, w_0 is the waist size, k is the wave number, $R(z)$ is the radius of curvature, and ϕ_G is the Gouy phase, as defined by Eqs. (3.83)–(3.85). Hermit polynomials of the first several orders are as follows:

$$H_0(x) = 1$$
$$H_1(x) = 2x$$
$$H_2(x) = 4x^2 - 2$$
$$H_3(x) = 8x^3 - 12x$$
$$H_4(x) = 16x^4 - 48x^2 + 12$$

Figure 3.17 plots the one-dimensional profile of Hermite–Gaussian mode for m up to the fourth order.

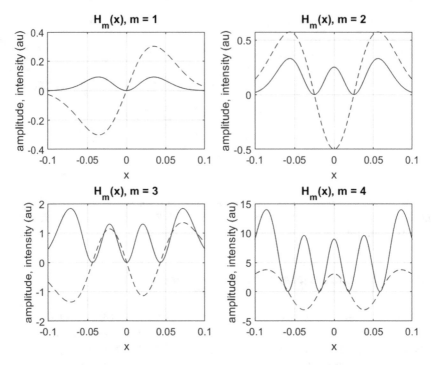

Fig. 3.17 Hermite–Gaussian beams of various orders; one-dimensional profile of amplitude (dashed lines) and intensity (solid lines)

Figure 3.18 illustrates the two-dimensional intensity profile of Hermite–Gaussian modes corresponding to Fig. 3.17. When $m = 1$ and $n = 0$, there are two intensity maxima horizontally, while there is only one maximum vertically. Likewise, other higher order modes exhibit maxima according to the mode numbers m and n. Figure 3.19 shows the intensity profile of the four modes used in Fig. 3.18 three-dimensionally.

Figure 3.20 shows symmetric ($m = n$) Hermite–Gaussian intensity profile for higher orders. Often a high-power laser having a symmetric resonator around the axis of the laser beam oscillate at a higher order Gaussian mode showing one of these patterns.

Orthogonality of Hermite polynomials

The Hermite polynomials of the above form $H_m(x)$ are orthogonal [41, 42] with respect to the weight function $exp(-x^2)$.

$$\int_{\infty}^{\infty} H_m(x)H_n(x)e^{-x^2}dx = \sqrt{\pi}2^n!\delta_{mn} \tag{3.103}$$

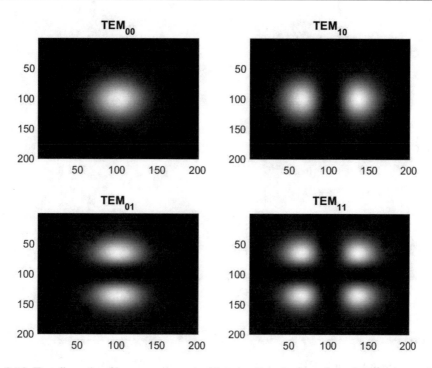

Fig. 3.18 Two-dimensional images representing Hermite–Gaussian beam intensity of various orders. Numbers on axes represent pixel positions

Here, δ_{mn} is the Kronecker delta. Consider expanding function $f(x)$ into Hermite–Gaussian modes.

$$f(x) = q_0 HG_0(x) + q_1 HG_1(x) + q_2 HG_2(x) + \cdots = \sum_{m=0}^{\infty} q_m HG_m(x) \qquad (3.104)$$

Here, HG_m is the mth mode of the Hermite–Gaussian series. We can use the above orthogonality to find the coefficient of the mth mode as follows:

$$\int_{-\infty}^{\infty} f(x) HG_m(x) dx = 0 + \cdots q_m \int_{-\infty}^{\infty} HG_m(x) HG_m(x) dx + 0 + \cdots \qquad (3.105)$$

$$q_m = \frac{\int_{-\infty}^{\infty} f(x) HG_m(x) dx}{\int_{-\infty}^{\infty} HG_m(x) HG_m(x) dx} \qquad (3.106)$$

The integral on the left-hand side of Eq. (3.105) is referred to as the overlap integral.

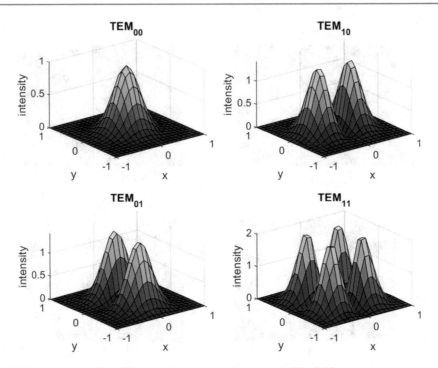

Fig. 3.19 Intensity profile of Hermite–Gaussian modes shown in Fig. 3.18

A high-power laser often outputs multi-mode beams consisting of a mixture of Hermite–Gaussian modes. From Eqs. (5.61) and (3.104), we can easily understand that the optical field of such a multi-mode laser output takes the following form:

$$E(t, x, y, z) = \sum_m \sum_n E_{m,n}$$

$$= E_0 \sum_m \sum_n \frac{w_0}{w(z)} H_m \left(\frac{\sqrt{2}x}{w(z)} \right) H_n \left(\frac{\sqrt{2}y}{w(z)} \right) exp \left[-\frac{x^2 + y^2}{w^2(z)} \right]$$

$$exp \left[-\frac{ik(x^2 + y^2)}{2R(z)} - ikz + i(m + n + 1)\phi_G \right] \quad (3.107)$$

We can find the coefficient of the mn mode via an integral similar to Eq. (3.105)

$$q_{mn} = \frac{\iint E(t, x, y, z) E'_{m,n} dx dy}{\iint E_{m,n} E'_{m,n} dx dy} \quad (3.108)$$

Here, $E'_{m,n}$ is the complex conjugate of $E_{m,n}$.

We can apply the above concept of expanding a laser mode into Hermite polynomials to the improvement of the beam quality. Having a large gain volume (see next section for lasers) contributes to high-power output but the resultant higher order modes compromise

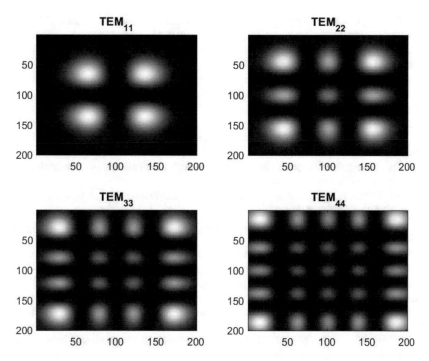

Fig. 3.20 Two-dimensional images representing Hermite–Gaussian beam intensity of symmetric higher orders. Numbers on axes represent pixel positions

the beam quality. Say, for instance, a high-power laser resonator oscillates in the TEM_{10} mode. As indicated by Fig. 3.19, there are two intensity peaks inside the cross-sectional area of the gain volume. As you can easily imagine, having two peaks, a laser beam does not focus well. The three-dimensional intensity profile in Fig. 3.19 indicates that each of the two intensity lobes resembles a TEM_{00} mode of smaller beam size. This observation indicates that it is possible to expand the lobe into a series of Hermit–Gaussian modes in such a way that the coefficient of the TEM_{00} mode is high. Buric et al. [43] demonstrated that by separating the two lobes of a TEM_{00} mode and coupling them into separate single-mode optical fiber nearly 94% of the power was converted to a TEM_{00} mode. The beam quality of a non-TEM_{00} mode laser is quantified with the parameter known as M^2 factor, or beam quality factor (M> 1) [30].

References

1. K. D. Bonin, V. V. Kresin, *Electric-dipole Polarizabilities of atoms, molecules, and clusters* (World Scientific, Singapore, New Jersey, 1997a)
2. *Dielectric Materials and Applications*, ed. by A. von Hippel (Artech House, Boston, London, 1995)

3. Dipole Moments, LibreTexts, https://chem.libretexts.org/Bookshelves/Physical_and_Theoretical_Chemistry_Textbook_Maps/Supplemental_Modules_(Physical_and_Theoretical_Chemistry)/Physical_Properties_of_Matter/Atomic_and_Molecular_Properties/Dipole_Moments (accessed on August 4, 2022)

4. K. D. Bonin, V. V. Kresin, *Electric-dipole Polarizabilities of atoms, molecules, and clusters* (World Scientific, Singapore, New Jersey, 1997b)

5. D. J. Griffiths, *Introduction to electrodynamics*. 3rd edn. (Prentice Hall, Upper Saddle River, NJ, USA, 1999) pp. 175 -179

6. G. D. Gillen, K. Gillen, S. Guha, *Light Propagation in Linear Optical Media* (CRC Press, Boca Raton, London, 2014)

7. 6.8 The Divergence Theorem, https://openstax.org/books/calculus-volume-3/pages/6-8-the-divergence-theorem (accessed on August 4, 2022)

8. D. J. Griffiths, *Introduction to electrodynamics*. 3rd edn. (Prentice Hall, Upper Saddle River, NJ, USA, 1999) pp. 269–273

9. D. J. Griffiths, *Introduction to electrodynamics*. 3rd edn. (Prentice Hall, Upper Saddle River, NJ, USA, 1999)

10. M. E. Thomas, *Optical Propagation in Linear Media* (Oxford Univ. Press, New York, 2006)

11. R. W. Boyd, *Nonlinear Optics*, 4th edn. (Academic Press, London, 2020)

12. D. L. Mills, *Nonlinear Optics*, (Springer-Verlag, Berlin, Heiderberg, 1991)

13. A. Yariv, *Optical Electronics*, 4th edn. (Saunders College Publishing, Philadelphia, 1991)

14. L. J. Wang, A. Kuzmich, A. Dogariu, Gain-assisted superluminal light propagation, Nature **406** 277–279, 2000

15. S. Yoshida, *Waves; Fundamental and dynamics* (Morgan & Claypool, San Rafael, CA, USA, IOP Publishing, Bristol, UK, 2017), p.3-7

16. Engineering Acoustics/Reflection and transmission of planar waves https://en.wikibooks.org/wiki/Engineering_Acoustics/Reflection_and_transmission_of_planar_waves (accessed on August 4, 2022)

17. A. Labeyrie, S. G. Lipson, P. Nisenson, *An Introduction to Optical Steller Interferometry* (Cambridge Univ. Press, Cambridge, UK, 2006) pp. 1-63

18. D. F. Buscher, *Practical Optical Interferometry, Imaging at Visible and Infrared Wavelengths* (Cambridge Univ. Press, Cambridge, UK, 2015)

19. M. Lucki, L. Bohac, R. Zeleny, *Fiber Optic and Free Space Michelson Interferometer — Principle and Practice*, InTech Open, https://doi.org/10.5772/57149,2014.

20. B. C. Barish, R. Weiss, LIGO and detection of Gravitational waves, Phys. Today **52** 44–50, 1999.

21. C. Sciammarella, Experimental Mechanics at the Nanometric Level, Strain **44**, 3-19, 2008.

22. E. Hecht, *Optics* 4th edn. (Addison Wesley, San Francisco, CA, USA, 2002) Chapter 8, pp. 325-384.

23. G. Chartier, *Introduction to Optics* (Springer, New York, 2005) pp 179–258

24. Y. Otani, 26 Birefringence Measurement, T. Yoshizawa, Ed. *Handbook of Optical Metrology*, 2nd, edn. O'Reilly https://www.oreilly.com/library/view/handbook-of-optical/9781466573598/xhtml/37_Chapter26.xhtml (accessed on August 7, 2022)

25. R. Pachotta, Waveplates, RP Photonics Encyclopedia, https://www.rp-photonics.com/waveplates.html (accessed on August 7, 2022)

26. Understanding Waveplates and Retarders, https://www.edmundoptics.com/knowledge-center/application-notes/optics/understanding-waveplates/#:%CB%9C:text=Fast%20Axis%20and%20Slow%20Axis,polarized%20along%20the%20slow%20axis (accessed on August 7, 2022)

27. Polarization Control with Optics, https://www.newport.com/n/polarization-control-with-optics#:%CB%9C:text=S%2Dpolarization%20refers%20to%20the,remaining%20figures %20of%20the%20section (accessed on August 7, 2022)

28. A. Yariv, *Introduction to Optical Electronics* (Holt, Rinehart and Winston, New York, USA, 1971)
29. J.-C. Nédélec, *Acoustic and Electromagnetic Equations* (Springer, New York, 2001) pp. 9-109
30. W. D. Kimura, *Electromagnetic Waves and Lasers*, 1st edn. (Morgan and Claypool, San Rafael, CA, USA, IOP Publishing, Bristol, UK, 2017) Chapter 1
31. R. Pachotta, Gaussian Beams, RP Photonics Encyclopedia, https://www.rp-photonics.com/gaussian_beams.html (accessed on August 7, 2022)
32. A. Talatinian, The analysis of the laser beam shape of the fundamental Gaussian mode by testing the numerical angular spectrum technique, Optics & Laser Technology, **118**, 75-83, 2019
33. M. R. Dennis, M. A. Alonso, Gaussian mode families from systems of rays, J. Phys. Photonics, **1**, (2), 025003, 2019.
34. R. Scheps, *Introduction to Laser Diode-Pumped Solid State Lasers* (SPIE Digital Library, 2002) https://doi.org/10.1117/3.2279412.ch2
35. A.E. Siegman, *Lasers* (University Science Books, Sausalito, CA, USA,1986)
36. A. Yariv, *Introduction to Optical Electronics* (Holt, Rinehart and Winston, New York, USA, 1971), Chap. 3
37. F. L. Pedrotti, L. M. Pedrotti, L. S. Pedrotti, *Introduction to Optics* 3rd edn. (Cambridge Univ. Press, 2018) pp. 593-597
38. A. Yariv, *Quantum Electronics* 3d edn. (New York: John Wiley & Sons, 1989), Chap. 6.
39. R. Pachotta, Hermite-Gaussian Modes, RP Photonics Encyclopedia, https://www.rp-photonics.com/hermite_gaussian_modes.html
40. H. J. Weber, G. B. Arfken, *Essential Mathematical Methods for Physicists* (Academic Press, Amsterdam, New York, 2003) pp. 638-662
41. 5.7.3: Hermite Polynomials are Orthogonal, LibreTexts, https://chem.libretexts.org/Bookshelves/Physical_and_Theoretical_Chemistry_Textbook_Maps/Physical_Chemistry_(LibreTexts)/05%3A_The_Harmonic_Oscillator_and_the_Rigid_Rotor/5.07%3A_Hermite_Polynomials_are_either_Even_or_Odd_Functions (accessed on August 7, 2022)
42. T. Alieva, M. J. Bastiaans, Mapping of Hermite-Gaussian modes in ABCD systems, Proc. **6027**, IOC20: Optical Information Processing; 60270C, 2006, https://doi.org/10.1117/12.667743
43. M. P. Buric J. Falk, S. D. Woodruff, Conversion of a TEM10 beam into two nearly Gaussian beams. Appl. Opt **49**, 739-744, 2010.

Properties of Light

4

4.1 Reflection and Refraction

4.1.1 Laws of Reflection and Refraction

In the preceding section, we discussed the reflection and transmission of waves when the incident wave is normal to the boundary plane. When a wave is incident obliquely to a boundary plane, the reflected and transmitted waves propagate in certain directions. These directions are represented by the angle of reflection and refraction, which is defined as the angle made by the propagation vector and a line normal to the boundary plane. We can use the laws of reflection and refraction to find the angles of reflection and refraction.

Figure 4.1 illustrates the reflection and refraction of a wave. Consider that a wave of wavelength λ is incident to a planar boundary. Let θ_i, θ_r, and θ_t be the angle of incidence, reflection, and refraction, respectively. Express the phase velocity of the wave in the medium on the incident wave side of the boundary, designated by 1, and that in the medium on the refracted wave side of the boundary, designated by 2, as follows:

$$v_1 = \frac{v_0}{n_1} \tag{4.1}$$

$$v_2 = \frac{v_0}{n_2} \tag{4.2}$$

Here, v_0 is the reference phase velocity, and n_1 and n_2 are the index of refraction for the respective media.

Since the frequency of the wave does not change at the boundary, the wavelengths inside medium 1 and 2 become as follows:

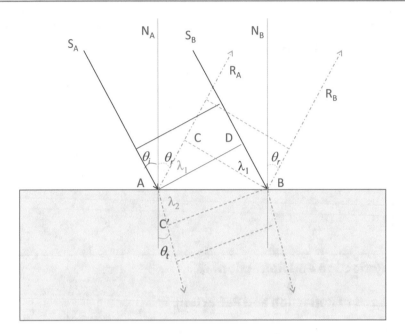

Fig. 4.1 Reflection and transmission of wave

$$\lambda_1 = \frac{\lambda_0}{n_1} \tag{4.3}$$

$$\lambda_2 = \frac{\lambda_0}{n_2} \tag{4.4}$$

Here, $\lambda_0 = v_0/\nu$ with ν being the frequency.

With the above information, we can find the relation between θ_i and θ_r, and θ_i and θ_t, respectively.

First consider the relation between θ_i and θ_r. In Fig. 4.1, solid lines $S_A A$ and $S_B B$ are two representative propagation vectors of the incident wave and the dashed lines connecting these two lines are two representative wavefronts that are apart by one wavelength. Dash-dotted lines AR_A and BR_B are two representative propagation vectors of the reflected wave. The dashed lines associated with AR_A and BR_B are also wavefronts one wavelength apart. Compare triangles ADB and ACB, who share side AB. The angle made by lines AD and AB is equal to the angle of incidence because both are a rectangle minus angle $N_A AD$. Similarly, the angle made by lines BC and AB is equal to the angle of reflection, θ_r. Since the incident and reflected waves are in medium 1, their wavelengths are λ_1 and equal to each other. Therefore, these two triangles are congruent. Consequently,

$$\theta_i = \theta_r \tag{4.5}$$

Relation (4.5) is the law of reflection.

Next, consider the relation between θ_i and θ_t by comparing triangles ADB and AC'B. Here, dash-dotted lines in medium 2 are two representative propagation vectors of the refracted wave, and dashed lines associated with these propagation vectors are two wavefronts apart by one wavelength λ_2. In this case, triangles ADB and AC'B are not congruent because AC' $\neq$ BD. However, the following relation is true:

$$\lambda_1 = \text{AB} \sin \theta_i \tag{4.6}$$

$$\lambda_2 = \text{AB} \sin \theta_t \tag{4.7}$$

Substitution of Eqs. (4.3) and (4.4) into Eqs. (4.6) and (4.7) leads to the following equation:

$$n_1 \sin \theta_i = n_2 \sin \theta_t \tag{4.8}$$

Relation (4.8) is the law of refraction or Snell's law [1, 2].

4.1.2 Coefficients of Reflection and Refraction

At the boundary of media having different indices of refraction, the following boundary conditions [4] hold. Consider Fig. 4.2.

$$E_{1\parallel} = E_{2\parallel} \tag{4.9}$$

$$\frac{B_{1\parallel}}{\mu_1} = \frac{B_{2\parallel}}{\mu_2} \tag{4.10}$$

Here, subscript $\parallel$ denotes the field components parallel to the boundary surface, and 1 and 2 denote the first and second mediums, respectively.

Boundary condition (4.9) is readily derived from Faraday's law (2.3). Application of Stokes' theorem [5] to Faraday's law yields the following equation:

$$\oint \mathbf{E} d\mathbf{l} = - \iint \frac{\partial \mathbf{B}}{\partial t} \cdot d\mathbf{S} \tag{4.11}$$

Consider a loop across the boundary plane in Fig. 4.2 to perform the closed line integral on the left-hand side of Eq. (4.11). The loop has width w and thickness h, parallel and perpendicular to the boundary plane, respectively. Since we consider Faraday's law on the boundary plane, the thickness of the loop, h, is null. Therefore, the left-hand side of Eq. (4.11) becomes as follows:

$$\oint \mathbf{E}d\mathbf{l} = \int_{z_0-h/2}^{0} E_{2z}dz + \int_{0}^{z_0+h/2} E_{1z}dz + \int_{x_0-w/2}^{x_0+w/2} E_{1x}dx$$

$$+ \int_{z_0+h/2}^{0} E_{1z}dz + \int_{0}^{z_0-h/2} E_{2z}dz + \int_{x_0+w/2}^{x_0-w/2} E_{2x}dx$$

$$= E_{1x} \int_{x_0-w/2}^{x_0+w/2} dx + E_{2x} \int_{x_0+w/2}^{x_0-w/2} dx$$

$$= E_{1x}w - E_{2x}w = (E_{1x}w - E_{2x}w)\, w \tag{4.12}$$

Here, z_0 and x_0 are the coordinate points of the middle of h and w, respectively. We take the line integral clockwise when viewed from the negative side of the y-axis. E_{1z} and E_{2z}, and E_{1x} and E_{2x} are the z and x-components of the electric field vector above and below the boundary plane. Since the thickness h is null, the magnetic flux is null on the right-hand side of Eq. (4.9). Thus, we obtain the following equation:

$$\oint \mathbf{E}d\mathbf{l} = (E_{1x} - E_{2x})w = (E_{1\parallel} - E_{2\parallel})w = 0 \tag{4.13}$$

Since the x-axis is parallel to the boundary plane, $E_{1x} = E_{1\parallel}$ and $E_{2x} = E_{2\parallel}$. Equation (4.13) leads to boundary condition (4.9), i.e., the parallel components of the electric field are continuous at a boundary.

Boundary condition (4.10) can be derived from Ampère's law (2.2). Consider in Fig. 4.2 surface current density $\mathbf{j}$ flowing on the boundary plane. Assuming that mediums 1 and 2 are linear (See Sect. 3.1.2) so that $\mathbf{D} = \epsilon\mathbf{E}$ and $\mathbf{B} = \mu\mathbf{H}$, and applying Stokes' theorem to Ampère's law we obtain the following equation.

$$\oint \mathbf{H}d\mathbf{l} = \iint \left(\frac{\partial \mathbf{D}}{\partial t} + \mathbf{j}\right) \cdot d\mathbf{S} \tag{4.14}$$

Here, $\mathbf{D}$ is the electric flux density vector and $\mathbf{H}$ is the auxiliary magnetic field. Similar to the above-argued electric field case, the first term on the right-hand side of Eq. (4.14) is null and the closed line integral on the left-hand side involves only the vector $\mathbf{H}$'s component

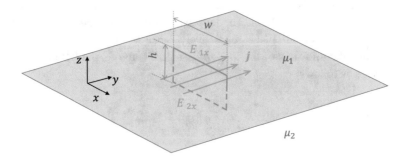

Fig. 4.2 Boundary condition for the electromagnetic field

parallel to the boundary plane. Thus, we obtain the following equation:

$$\oint \mathbf{H}d\mathbf{l} = H_{1x}w - H_{2x}w = (H_{1p} - H_{2p})w = \left(\frac{B_{1p}}{\mu_1} - \frac{B_{2p}}{\mu_2}\right)w = J \qquad (4.15)$$

Here, J is the total surface current enclosed by the $w \times h$ loop shown in Fig. 4.2. When no surface current flows, $J = 0$ and Eq. (4.15) leads to boundary condition (4.10).

While it is convenient to use the boundary plane to discuss the boundary condition, it is customary to use the p-polarization [6] and s-polarization [6] discussed under section "polarization" (also see Fig. 3.12). In the following section, we discuss the reflection and transmission of light waves with an s-polarized electric field and s-polarized magnetic field. The former is referred to as a TE (Transverse Electric) wave [8–10] and the latter a TM (Transverse Magnetic) wave [8, 10].

Reflection and transmission coefficients of E_s wave

In Fig. 4.3, consider the electric field is perpendicular to the plane of incidence E_s. In this case, the entire electric field vectors and the horizontal components of the magnetic vectors are parallel to the boundary plane. Therefore, the boundary conditions (4.9) and (4.10) become as follows:

$$E_i + E_r = E_t \qquad (4.16)$$

$$-\frac{B_i}{\mu_1}\cos\theta_i + \frac{B_r}{\mu_1}\cos\theta_i = -\frac{B_t}{\mu_2}\cos\theta_t \qquad (4.17)$$

Here, $-B_i\cos\theta_i$, $B_r\cos\theta_r$ and $-B_t\cos\theta_t$ are the magnetic vector components parallel to the boundary plane. Note that all these magnetic vectors are parallel to the plane of incidence, i.e., s wave (See Fig. 3.12).

Substituting $E = cB$ (2.75) into Eq. (4.17) and using the speed of light (2.25), we can rewrite Eq. (4.17) as follows:

$$-\sqrt{\frac{\epsilon_1}{\mu_1}}E_i\cos\theta_i + \sqrt{\frac{\epsilon_1}{\mu_1}}E_r\cos\theta_i = -\sqrt{\frac{\epsilon_2}{\mu_2}}E_t\cos\theta_t \qquad (4.18)$$

Fig. 4.3 Reflection and transmission of light when electric field is s-polarized and magnetic filed is p-polarized

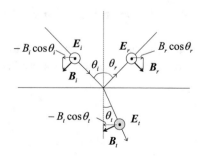

Solving Eqs. (4.16) and (4.18) for E_r and E_t, we obtain the amplitude reflection and transmission coefficient for the E_s case.

$$r_s \equiv \frac{E_r}{E_i} = \frac{\sqrt{\epsilon_1/\mu_1} \cos \theta_i - \sqrt{\epsilon_2/\mu_2} \cos \theta_t}{\sqrt{\epsilon_1/\mu_1} \cos \theta_i + \sqrt{\epsilon_2/\mu_2} \cos \theta_t} = \frac{z_2 \cos \theta_i - z_1 \cos \theta_t}{z_2 \cos \theta_i + z_1 \cos \theta_t} \tag{4.19}$$

$$t_s \equiv \frac{E_t}{E_i} = \frac{2\sqrt{\epsilon_1/\mu_1} \cos \theta_i}{\sqrt{\epsilon_1/\mu_1} \cos \theta_i + \sqrt{\epsilon_2/\mu_2} \cos \theta_t} = \frac{2z_2 \cos \theta_i}{z_2 \cos \theta_i + z_1 \cos \theta_t} \tag{4.20}$$

Here, z_1 and z_2 are the specific impedance [11, 12] defined by Eq. (3.54) for the electromagnetic wave. Suffix "1" and "2" denote the medium on the incident light side and transmitted light side, respectively. r_s and t_s are referred to as the amplitude coefficient of reflection and transmission for the E_s wave as they express the ratio of the corresponding amplitude. The suffix s is used to mean that the electric field of the light wave is s-polarized. Since the electromagnetic wave is a transverse wave, if the electric field is s-polarized, the magnetic field is p-polarized (Fig. 3.12). It is conventional to use the electric field's polarization to label an electromagnetic wave's polarization. Equations (4.19) and (4.20) are known as Fresnel equations [13] for s-polarized waves.

For practical uses, it is convenient to express the Fresnel equations using the index of refraction. From the speed of light expression (2.25) and the index of refraction (Eq. (4.1)),

$$c = \frac{c_0}{n} = \frac{1}{\sqrt{\epsilon\mu}} \tag{4.21}$$

where c_0 is the speed of light in free space. From Eqs. (3.54) and (4.21), we find

$$z = \sqrt{\frac{\mu}{\epsilon}} = \frac{c_0\mu}{n} \tag{4.22}$$

Using Eq. (4.22) and $\mu_1 \cong \mu_2$, we can write Eqs. (4.19) and (4.20) as follows:

$$r_s = \frac{E_r}{E_i} = \frac{\frac{c_0\mu_2}{n_2} \cos \theta_i - \frac{c_0\mu_1}{n_1} \cos \theta_t}{\frac{c_0\mu_2}{n_2} \cos \theta_i + \frac{c_0\mu_1}{n_1} \cos \theta_t} \cong \frac{n_1 \cos \theta_i - n_2 \cos \theta_t}{n_1 \cos \theta_i + n_2 \cos \theta_t} \tag{4.23}$$

$$t_s = \frac{E_t}{E_i} = \frac{2\frac{c_0\mu_2}{n_2} \cos \theta_i}{\frac{c_0\mu_2}{n_2} \cos \theta_i + \frac{c_0\mu_1}{n_1} \cos \theta_t} \cong \frac{2n_1 \cos \theta_i}{n_1 \cos \theta_i + n_2 \cos \theta_t} \tag{4.24}$$

Reflection and transmission coefficients of E_p wave

When the electric field is parallel to the plane of incidence, the boundary conditions (4.9) and (4.10) become as follows; see Fig. 4.4.

$$E_i \cos \theta_i - E_r \cos \theta_i = E_t \cos \theta_t \tag{4.25}$$

$$\frac{B_i}{\mu_1} + \frac{B_r}{\mu_1} = \frac{B_t}{\mu_2} \tag{4.26}$$

Fig. 4.4 Reflection and transmission of light when electric field is p-polarized and magnetic filed is s-polarized

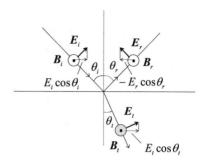

Here, Eq. (4.25) represents the condition for the components of the electric field parallel to the boundary plane. This time the entire magnetic field vectors are parallel to the boundary plane. Equation (4.26) represents the conditions for the magnetic fields.

Repeating the same argument as the E_s case, we can derive the following expressions for the amplitude reflection and transmission coefficient of the E_p case:

$$r_p = \frac{\sqrt{\epsilon_2/\mu_2}\cos\theta_i - \sqrt{\epsilon_1/\mu_1}\cos\theta_t}{\sqrt{\epsilon_2/\mu_2}\cos\theta_i + \sqrt{\epsilon_1/\mu_1}\cos\theta_t} = \frac{z_1\cos\theta_i - z_2\cos\theta_t}{z_1\cos\theta_i + z_2\cos\theta_t} \tag{4.27}$$

$$t_p = \frac{2\sqrt{\epsilon_1/\mu_1}\cos\theta_i}{\sqrt{\epsilon_2/\mu_2}\cos\theta_i + \sqrt{\epsilon_1/\mu_1}\cos\theta_t} = \frac{2z_2\cos\theta_i}{z_1\cos\theta_i + z_2\cos\theta_t} \tag{4.28}$$

Equations (4.27) and (4.28) are known as Fresnel equations [13] for p-polarized waves. Here, the polarization is for the electric field as is the case of s-polarized waves; if the polarization of E is "p", the polarization of B is "s".

Similarly to the E_s wave case, we can express reflection and transmission coefficients using the index of refraction.

$$r_p = \frac{\frac{c_0\mu_1}{n_1}\cos\theta_i - \frac{c_0\mu_2}{n_2}\cos\theta_t}{\frac{c_0\mu_1}{n_1}\cos\theta_i + \frac{c_0\mu_2}{n_2}\cos\theta_t Z} \simeq \frac{n_2\cos\theta_i - n_1\cos\theta_t}{n_2\cos\theta_i + n_1\cos\theta_t} \tag{4.29}$$

$$t_p = \frac{2\frac{c_0\mu_2}{n_2}\cos\theta_i}{\frac{c_0\mu_1}{n_1}\cos\theta_i + \frac{c_0\mu_2}{n_2}\cos\theta_t} \simeq \frac{2n_1\cos\theta_i}{n_2\cos\theta_i + n_1\cos\theta_t} \tag{4.30}$$

Power reflectance and transmittance

So far we discussed the reflection and transmission coefficients for amplitude. Here, we consider the reflection and transmission of power. The reflection and transmission coefficients for power are called reflectance (R) and transmittance (T).

From the relation $S_0 = E_0^2/z$ (Eq. (3.51)), you may think that $R = r^2$ and $T = t^2$. While the former is true the latter is not true. This is because Eq. (3.51) is true for a light wave propagating in a uniform medium. Therefore, Eq. (3.51) is not applicable to the transmission that involves two media having different refractive indexes. In order to address the case involving two refractive indices, we need to calculate the Poynting vector for the incident, reflected, and transmitted waves, respectively. From the relation that the ratio of the amplitude of the electric field over the amplitude of the magnetic field is the speed of light, we obtain the following equations:

$$B_i = \frac{E_i}{c_1}, \quad B_r = \frac{E_r}{c_1}, \quad B_t = \frac{E_t}{c_2} \tag{4.31}$$

$$S_i = \frac{E_i B_i \cos\theta_i}{\mu_1}, \quad S_r = \frac{E_r B_r \cos\theta_i}{\mu_1}, \quad S_t = \frac{E_t B_t \cos\theta_t}{\mu_2} \tag{4.32}$$

Here, the subscripts "i", "r", and "t" represent "incident", "reflected", and "transmitted", and S is the amplitude of the Poynting vector. Substitution of Eq. (4.31) into Eq. (4.32) yields the following expressions:

$$S_i = \frac{E_i E_i \cos\theta_i}{\mu_1 c_1} = \frac{E_i^2 \cos\theta_i}{z_1} \tag{4.33}$$

$$S_r = \frac{E_r E_r \cos\theta_i}{\mu_1 c_1} = \frac{E_r^2 \cos\theta_i}{z_1} \tag{4.34}$$

$$S_t = \frac{E_t E_t \cos\theta_t}{\mu_2 c_2} = \frac{E_t^2 \cos\theta_t}{z_2} \tag{4.35}$$

From Eqs. (4.33), (4.34) and (4.35), we find the reflectance and transmittance as follows:

$$R = \frac{S_r}{S_i} = \frac{E_r^2 \cos\theta_i / z_1}{E_i^2 \cos\theta_i / z_1} = r^2 \tag{4.36}$$

$$T = \frac{S_t}{S_i} = \frac{E_t^2 \cos\theta_t / z_2}{E_i^2 \cos\theta_i / z_1} = t^2 \frac{z_1 \cos\theta_t}{z_2 \cos\theta_i} \tag{4.37}$$

For the E_s wave case, substituting Eq. (4.19) into (4.36), and Eq. (4.20) into (4.37), we obtain the following equations:

$$R_s = \frac{(z_2 \cos\theta_i - z_1 \cos\theta_t)^2}{(z_2 \cos\theta_i + z_1 \cos\theta_t)^2} \tag{4.38}$$

$$T_s = \frac{(2z_2 \cos\theta_i)^2}{(z_2 \cos\theta_i + z_1 \cos\theta_t)^2} \frac{z_1 \cos\theta_t}{z_2 \cos\theta_i} = \frac{4z_1 z_2 \cos\theta_i \cos\theta_t}{(z_2 \cos\theta_i + z_1 \cos\theta_t)^2} \tag{4.39}$$

Apparently, Eqs. (4.38) and (4.39) indicate $R_s + T_s = 1$. It follows that

$$S_r + S_t = R_s S_i + T_s S_i = (R_s + T_s)S_i = S_i \tag{4.40}$$

Equation (4.40) represents that the optical energy is conserved at the boundary.

With the same procedure as the E_s case, we can derive the power reflectance and transmittance for the p wave case as follows:

$$R_p = \frac{(z_1 \cos \theta_i - z_2 \cos \theta_t)^2}{(z_1 \cos \theta_i + z_2 \cos \theta_t)^2} \tag{4.41}$$

$$T_p = \frac{(2z_2 \cos \theta_i)^2}{(z_1 \cos \theta_i + z_2 \cos \theta_t)^2} \frac{z_1 \cos \theta_t}{z_2 \cos \theta_i} = \frac{4z_1 z_2 \cos \theta_i \cos \theta_t}{(z_1 \cos \theta_i + z_2 \cos \theta_t)^2} \tag{4.42}$$

Again, $R_p + T_p = 1$ holds, indicating the conservation of the optical energy.

4.1.3 External and Internal Reflection

External reflection

The reflection that occurs when a wave travels from a medium of the lower refractive index (n_1) to a higher ($n_2 > n_1$) is referred to as external reflection [3], e.g., from air to glass. According to Snell's law (4.8), the angle of refraction is given as follows:

$$\theta_t = \sin^{-1}\left(\frac{n_1}{n_2} \sin \theta_i\right) \tag{4.43}$$

Here, θ_i is the angle of incidence, n_1 and n_2 are the index of refraction on the incidence side and transmitted side, respectively. Since $n_2 > n_1$, the value inside the parenthesis on the right-hand side of Eq. (4.43) is less than unity, regardless of the value of θ_i (because $|\sin \theta_i| \leq 1$). Hence, we can always find the angle of refraction, meaning that transmission necessarily occurs.

Figure 4.5a plots the amplitude coefficients of reflection and transmission for air to glass. Notice that the reflection and transmission coefficients are significantly different between the s-polarization and p-polarization.

Internal reflection

When a wave is incident to a medium whose index of refraction is lower ($n_1 > n_2$), the resultant reflection is called internal reflection. Consider increasing the angle of incidence under this condition. According to Eq. (4.8), the angle of refraction θ_t reaches 90° earlier than θ_i. Call the angle of incidence that makes the angle of refraction equal to 90° the critical angle θ_c. From Eq. (4.8),

$$\sin \theta_c = \frac{n_2}{n_1} \tag{4.44}$$

For an angle of incidence greater than the critical angle ($\theta_i > \theta_c$), the value inside the parenthesis on the right-hand side of Eq. (4.43) becomes greater than unity.

$$\left(\frac{n_1}{n_2} \sin \theta_i\right) > \left(\frac{n_1}{n_2} \sin \theta_c\right) = \left(\frac{n_1}{n_2} \frac{n_2}{n_1}\right) = 1 \tag{4.45}$$

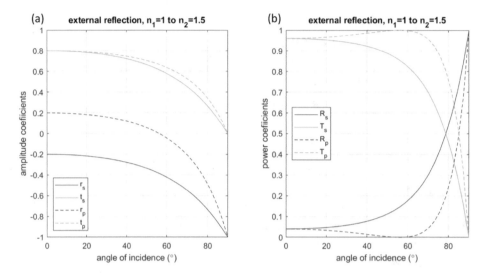

Fig. 4.5 Reflection and transmission of s- and p-polarized waves when they are incident to glass from air

Since the sign function cannot be greater than unity, it becomes impossible to find θ_i under condition (4.45). Physically, this means that transmission does not occur, or the incident light is totally reflected. This phenomenon is known as total reflection. Since the sine function monotonically increases in the first quadrant, total reflection occurs at any angle of incidence greater than the critical angle.

Figure 4.6 plots the amplitude and power coefficients of reflection and transmission when an s- or p-polarized light is incident from glass to air. The boundary condition is the same as Fig. 4.5 except that the incident and transmitted sides are switched. Using Eq. (4.43) with the refractive index of unity for air and 1.5 for glass, we can find the critical angle for this case to be 41.8° as indicated with an arrow in Fig. 4.6a. In Fig. 4.6b, we can see that above this critical angle the transmittance is null. Figure 4.6a indicates that the amplitude coefficients of transmission are greater than unity. You may think it strange because if you square these amplitude transmission coefficients the result is greater than unity and therefore the transmitted power is greater than the incident power. However, this argument is not correct because transmittance is not the square of the amplitude transmission coefficient. Remember Eq. (4.42). The factor $(z_1 \cos \theta_t)/(z_2 \cos \theta_i)$ makes the optical energy be conserved at the boundary. (Hecht published an interesting article about this issue [14]).

One of the important applications of internal reflection involves a type of polarizer known as Glan-type polarizing prisms [15, 16]. They are used to separate a linearly polarized light from unpolarized or poorly polarized light. Figure 4.7 illustrates a typical polarizing prism called the Glan–Foucault prism [16]. It consists of two identical pieces of birefringent material [18, 19], typically calcite [17, 20] (calcium carbonate, $CaCO_3$) separated by a thin

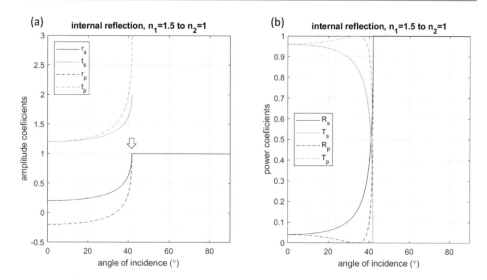

Fig. 4.6 Reflection and transmission of s- and p-polarized waves when they are incident to air from glass

Fig. 4.7 Glan–Foucault prism polarizer

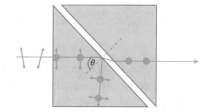

air gap. The incident unpolarized light enters from the left surface with normal incidence, as shown, and reaches the air gap where it undergoes internal reflection. Calcite is a birefringent crystal and therefore the incident beam is decomposed into ordinary and extraordinary waves. The crystal orientation is adjusted (when the prism is manufactured) so that the extraordinary wave is s-polarized and the ordinary wave is p-polarized. By the mechanism discussed in the next paragraph, only the s-polarized wave comes out of the prism from the right surface.

The separation of the s-wave is made as follows. Calcite's refractive index for the ordinary wave is $n_o = 1.6584$ and that of the extraordinary wave is $n_e = 1.4864$ at wavelength $\lambda = 589.3$ nm [17]. Since the angle of incidence is $90°$, the ordinary and extraordinary waves propagate along the same path until they reach the air gap. However, because of the difference in the refractive index, their critical angles are different from each other. Figure 4.8 plots the amplitude coefficient of reflectivity for the internal reflection of the s and p-polarization. Notice that at the angle of incidence of approximately $37°$ the p-polarization enters the total reflection regime but the s-polarization is still in the partial reflection regime. Therefore, if the angle of incidence at the calcite-to-air surface is slightly greater than this p-wave's

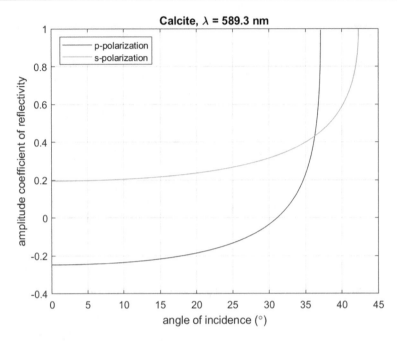

Fig. 4.8 Amplitude coefficient of reflectivity for *p*- and *s*-polarization in calcite

critical angle, the ordinary wave is totally reflected and does not enter the air gap. On the other hand, the extraordinary wave is partially reflected and exits the prism from the right surface. In this fashion, the light wave transmitted through the prism is *s*-polarized. Note that the reflected beam consists of *p*- and *s*-polarized waves.

The second crystal on the right side of the air gap does not contribute to the separation of the *s*-polarization. It is used to make the transmitted beam parallel to the incident beam. Because the angle of incidence to the air gap is not normal, the *s*-polarized beam is bent at the calcite-to-air boundary. However, when it enters the second crystal, the *s*-polarized beam undergoes the same refraction, and, therefore inside the second crystal, its path is again parallel to the incident beam. Since the incidence to the final surface is normal, the *s*-polarize beam comes out of the prism without being bent. There is a small lateral shift due to the path inside the air gap. However, since the air gap is small usually this lateral shift is negligible.

4.2   Interference

We discussed interference of waves in association with interferometry in Sect. 3.2.3. Interferometry makes use of interference to obtain information, and therefore, we want to enhance interference in an interferometer. Interference can be harmful in various situations where we want to suppress it. Here, we discuss interference in more general including practical problems related to interferometry.

Interference occurs when two or more waves occupy the same spatial location at the same time. Consider two waves $E_1 e^{i(\omega_1 t - \mathbf{k}_1 \cdot \mathbf{r})}$ and $E_2 e^{i(\omega_2 t - \mathbf{k}_2 \cdot \mathbf{r} + \phi)}$. Assume that the location where they are being superposed is $\mathbf{r} = 0$. (We can use any value of $\mathbf{r}$ and, by setting $\mathbf{r} = 0$, we do not lose the generality of the argument.) In this case, we can express the total intensity at this location as follows:

$$
\begin{aligned}
I_{sum} &= (E_1 e^{i\omega_1 t} + E_2 e^{i(\omega_2 t + \phi)})(E_1 e^{-i\omega_1 t} + E_2 e^{-(i\omega_2 t + \phi)}) \\
&= (E_1^2 + E_2^2) + 2E_1 E_2 \cos((\omega_2 - \omega_1)t + \phi)
\end{aligned}
\tag{4.46}
$$

Equation (4.46) indicates that the total intensity fluctuates at the differential frequency $\omega_2 - \omega_1$. Since interferometry detects the argument of the cosine function as the phase difference (between the two interfering waves) to analyze, this fluctuation is not favorable. If the differential frequency $\omega_2 - \omega_1$ is too high for the detector to resolve, the effect of interference is undetectable. If the frequencies ω_1 and ω_2 are close to each other, the fluctuation frequency can be detectable; the detector signal varies with time, as the $\cos(\omega_2 - \omega_1)t$ term oscillates in the range of -1 to 1. This type of fluctuation is known as beat [21]. For interferometry, it is essential that the interfering waves have the same frequency so that the cosine function solely depends on the phase difference phi that contains the information of interest.

When $\omega_1 = \omega_2$, Eq. (4.46) becomes as follows:

$$
I_{sum} = (E_1^2 + E_2^2) + 2E_1 E_2 \cos\phi
\tag{4.47}
$$

The total intensity has a dependence on ϕ, where it is at the maximum when $\phi = 2N\pi$ and at the minimum when $\phi = (2N + 1)\pi$ (N is an integer). Here, the maximum and minimum values depend on the relative value of the amplitude E_1 and E_2 as illustrated by Fig. 4.9. Notice that the maximum is the highest and the minimum is the lowest when the two amplitudes are equal to each other. As the amplitude ratio decreases, the maximum intensity decreases, and the minimum intensity increases. We say that the contrast between the maximum and minimum decreases as the ratio reduces.

This contrast is measured by a parameter known as visibility V.

$$
V = \frac{I_{max} - I_{min}}{I_{max} + I_{min}}
\tag{4.48}
$$

Notice that when $E_1 = E_2$, the visibility takes the highest value of $V = 1$ because $I_{max} = 1$ and $I_{min} = 0$. With the decrease in the ratio E_2/E_1, the visibility decreases approaching $V = 0$ when $E_2/E_1 = 0$. In this limit, Eq. (4.47) indicates $I_{max} = I_{min} = E_1^2$, and therefore, from Eq. (4.48), we find $V = 0$. In these arguments, E_1 and E_2 are interchangeable.

Naturally, high visibility is advantageous for interferometry. When interference compromises the quality of data, low visibility is favorable. Below, we discuss several practical issues related to interference.

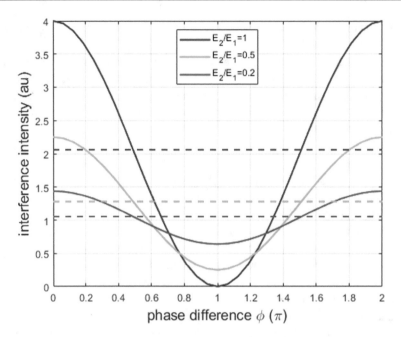

Fig. 4.9 Total intensity as a function of phase difference ϕ for three representative ratios of the amplitude. The dashed line is the average intensity for each ratio

4.2.1 Low Visibility in Interferometry

Sometimes, although we think we configure the interferometer correctly, the visibility is lower than expected. In this section, we discuss several factors that affect visibility using the interferometer configuration shown in Fig. 4.10.

Figure 4.10a illustrates a Michelson-type interferometer [25] configured to analyze the surface profile of a coated specimen [26]. The end mirror for the reference path is a usual reflector and the other end mirror is the specimen. The specimen is slightly tilted from the normal incidence so that the phase variation is visualized as the spatial variation of the grayscale. Here, the minimum phase difference corresponds to the lowest grayscale and the maximum phase difference to the highest grayscale. Thus, ideally, the phase of 0 rad corresponds to a completely dark pixel and $(1/2 + N)\pi$ rad to the brightest pixel. The former is called the dark fringe, and the latter the bright fringe. If the interference pattern has visibility of unity and the grayscale of the digital camera is 0–255 (8 bits), the grayscale of the dark fringe is 0 and that of the bright fringe is 255. Figure 4.10b and c are sample fringe patterns obtained by this interferometer setup. It is clear that fringe pattern (c) has significantly higher visibility than (b). Below, we will discuss typical factors that lower the fringe visibility.

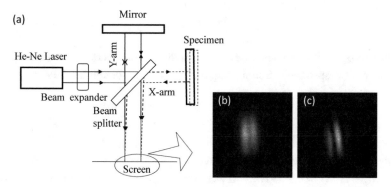

Fig. 4.10 Michelson-type interferometer for surface analysis. The dark (black) and bright (white) fringes appear in the central region of images **b** and **c**. The contrast of **b** is lower than **c**

Inbalanced amplitude

According to Fig. (4.9), the visibility is maximized when the interfering waves have the same amplitude. The amplitude of the interfering laser beams is different from each other for various reasons. Common reasons for this imbalance include an imbalance beam splitter (not splitting 50%), and a difference in the reflectivity between the two end mirrors. In the case of Fig. 4.10 configuration, the latter is the case. The vertical interferometer arm uses a 100% optical reflector and the other arm uses the specimen surface for the end mirror. Since the reflectivity of the specimen surface is lower than 100%, we need to match the reflection from the optical reflector accordingly to achieve high visibility.

Polarization

Another common cause for low visibility is that the polarization [27] of the interfering lights is not aligned. Interference is a superposition of the electric field of the interfering light waves, and therefore, if the polarization of one wave is not aligned to the other, only the parallel components interfere. Consequently, even if the interfering waves have the same intensity, the amplitude of the interfering portion is not balanced. Some beam splitters alter the polarization. It is important to choose the correct beam splitter. The management of the polarization of the laser source is important as well.

Coherence

Also, a common cause for poor performance is that the optical path difference of the interfering beams exceeds the coherent length of the laser. The coherent length depends on the monochromaticity of the lasing emission [28], and the resonator modes. One way to define the coherent length l_c is to set the phase difference resulting from the spectral width of the laser source equals 2π.

$$(k_1 - k_2)l_c = 2\pi \left(\frac{1}{\lambda} - \frac{1}{\lambda + \Delta\lambda} \right) l_c = 2\pi \frac{\Delta\lambda}{\lambda(\lambda + \Delta\lambda)} l_c \approx 2\pi \frac{\Delta\lambda}{\lambda^2} l_c = 2\pi \qquad (4.49)$$

Here, k_1 and k_2 are the wave number corresponding to the shortest and the longest end of the wavelength, respectively. Solving Eq. (4.49) for l_c, we obtain the following expression for l_c:

$$l_c = \frac{\lambda^2}{\Delta\lambda} = \lambda \frac{\lambda}{\Delta\lambda} = \frac{c}{\nu} \frac{\nu}{\Delta\nu} = \frac{c}{\Delta\nu} \qquad (4.50)$$

Here, ν and c are the frequency and the speed of light. In going through the third equal sign in Eq. (4.50), we use $c = (\nu - \Delta\nu)(\lambda + \Delta\lambda) = \nu\lambda + (\nu\Delta\lambda - \Delta\nu\lambda) - \Delta\nu\Delta\lambda \approx \nu\lambda + (\nu\Delta\lambda - \Delta\nu\lambda) = c + (\nu\Delta\lambda - \Delta\nu\lambda)$. The coherent length of a typical helium–neon laser oscillating at 632.8 nm is approximately 30 cm. It is a good idea to check the optical path of the respective beams from the beam splitter to the point where the two beams are recombined for interference. In the case of the Michelson interferometer shown in Fig. 3.4, for example, the optical path difference is twice the difference between the subject path l_s and reference path l_r (because the two beams are recombined after the round trip of each path). It is important that $2|l_s - lr|$ is within the same order as the coherent length of the laser used for the optical source.

Instability

The instability of the interferometer's optical path length can lower the visibility. When the interference becomes unstable, the relative phase difference fluctuates. In the case of the fringe patterns shown in Fig. 4.10, instability causes the dark and bright fringes to oscillate back and forth perpendicular to the fringe lines, resulting in a reduction of visibility. The study using this setup utilizes this visibility reduction to probe the characterize the coating strength on the specimen surface [26].

Among various causes of instability, mechanical vibration, such as the seismic (floor) motion, and temperature fluctuation of the air in the optical paths are commonly observed ones. As for the mechanical vibration, if the entire interferometer vibrates commonly to the two interfering paths, it does not significantly affect the fringe pattern. Noise of this sort is referred to as common noise. On the other hand, a vibration that causes the relative optical path length to fluctuate is usually a serious problem. This disturbance is referred to as differential noise. A vibration of the beam splitter is a possible source of differential noise. It directly alters the differential optical path length and often significantly compromises visibility.

Another common cause of interferometer instability is the temperature fluctuation of the medium on the optical paths. When the interferometric optical paths are in the air, for instance, the temperature dependence of the refractive index can have a significant influence. The temperature coefficient of the refractive index of air is of the order of 10^{-6} /K at room temperature for the helium–neon wavelength (632.8 nm). Since the total optical path length for a physical length of l m is $l/c = l/(c_0/n) = n(l/c_0)$ (c and c_0 are speed of light in air and vacuum), the change in optical path length for a 20 cm physical length due to

0.1 K of temperature fluctuation is $1 \times 10^{-6} \times 20 \times 0.1 = 1 \times 10^{-6}$ cm = 20 nm. This is 20/632.8=3.2% of the helium–neon laser's wavelength. This means that for a 10 cm long Michelson interferometer placed in air, if the air temperature of one interferometric path becomes higher or lower than the other path, the relative optical path length behind the beam splitter changes by 3.2 % of the period. This phase shift increases in proportion to the physical path length and the differential temperature change.

4.2.2 Interference of Multiple Light Rays and Speckles

Interference occurs when multiple waves are superposed with one another. We can easily understand the formation of interference with multiple waves by considering the following expression of the total amplitude:

$$A_{total} = A_1 e^{i\theta_1} + A_2 e^{i\theta_2} + \cdots + A_1 e^{i\theta_k} = A e^{i\theta} \tag{4.51}$$

Equation (4.51) indicates that no matter how many waves may be superposed, the amplitude of the total (summed) optical field can be expressed with a single pair of amplitude and phase (A, θ). Figure 4.11 illustrates the situation schematically.

The fringe images Fig. 4.10b and c show a number of bright and dark spots over the entire area. These spots, referred to as speckles [29], are formed by a coherent superposition of a number of light rays reflected from the object. When the object's surface is rough at the level of the wavelength, reflection becomes diffuse. Consequently, at a given spot of the detecting device such as a digital camera, a number of light rays originating from various points on the object's surface are superposed. According to the argument made with Fig. 4.11, the superposed light has a certain phase. Depending on this phase, the intensity of the spot varies from the darkest to the brightest. Since this phase is randomly distributed over

Fig. 4.11 Superposition of multiple waves

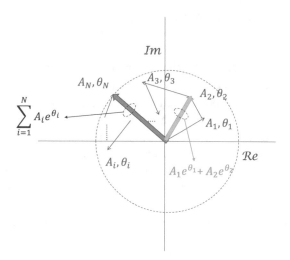

Fig. 4.12 Schematic
representation of speckles
formed by diffuse reflection

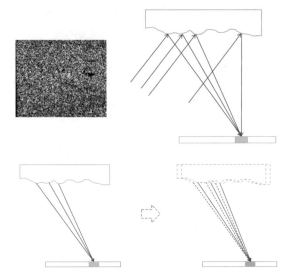

Fig. 4.13 Principle of speckle
metrology. The optical phase at
the pixel is altered by the
displacement

the surface of the image plane, a random pattern of bright and dark spots is formed. Figure 4.12 illustrates the situation schematically where light rays reflected at various points on the rough surface enter the same pixel of the imaging device.

For the image of a certain object, the speckle is usually considered as noise and called speckle noise. However, since each speckle has a definite phase, we can use speckles for metrology. The metrological technique is referred to as speckle metrology and is widely applied in deformation analysis [30–32]. Figure 4.13 illustrates the principle of displacement measurement based on the speckle metrology. Consider that rays of reflected light are captured by a pixel. Suppose that the object that reflects the rays is displaced to the right. Here the displacement is so small that we can assume that the image of the reflection remains the same. However, the distance from the light source to the object changes in proportion to the displacement, and hence, the phase of the light captured by the pixel changes accordingly. Therefore, if the grayscale value of this pixel before the displacement is subtracted from that after the displacement, the answer is non-zero, except for one condition that the phase change associated with the displacement is an integer multiple of 2π. The area that meets this condition shows a dark spot.

When the displacement is not uniform over the surface, i.e., the object undergoes deformation, the parts of the subtracted image where the above phase condition holds become dark. Figure 4.14 demonstrates this mechanism. Here, the "before" and "after" deformation images have a number of speckles. Since the deformation is so small, the speckle patterns appear unchanged between the "before" and "after" images. However, the phase change at all the points is different ranging over several periods. Consequently, the subtracted image exhibits seven dark lines. The pixels falling in these dark lines experience phase changes of integer multiples of 2π due to the deformation.

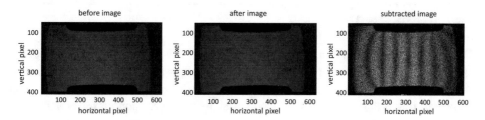

Fig. 4.14 Fringe formation by deformation of object surface

4.2.3 Unwanted Interference

Often interferometric fringe images like Fig. 4.14 are compromised by unwanted interference. Figure 4.15 shows fringe images similar to Fig. 4.14 for a notched specimen. Fringe patterns (a), (b), (c) are formed as the tip of the notch moves to the right. The linear fringes enclosed by the triangular dashed line represent the strain field ahead of the notch tip. It is seen that these fringe patterns move to the right with the notch tip. On the other hand, concentric circular fringes enclosed by the elliptical dashed line stay at the same location on the specimen in (a)–(c). Obviously, these circular fringes are formed by the interferometric setup and are unwanted. It is likely that these unwanted fringes result from multiple reflections inside the objective lens used to illuminate the specimen. The laser light is reflected by the front and rear surfaces of the multiple lenses placed inside the objective lens. Sometimes those ghost fringes confuse the signal fringes and should be avoided as much as possible. Often better alignment of lenses to the optical axis solves this problem.

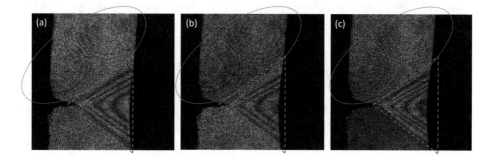

Fig. 4.15 Ghost circular fringes appear in signal fringes representing near notch strain field

4.3   Dispersion

Dispersion is the phenomenon observed when the material's property is such that the phase velocity depends on the frequency. The frequency of a light source has some width (called the bandwidth) due to some mechanism of spectral broadening. When a wave with a certain bandwidth travels through a dispersive medium, all frequency components travel at different phase velocities. This causes two practical effects. First, the shape of the wave changes. Second, the phase velocity becomes less meaningful as compared with the case when the wave does not show dispersion. In this case, the group velocity is used.

4.3.1   Group Velocity

The group velocity v_g can be related to the phase velocity v_p as follows:

$$v_p(k) = \frac{\omega}{k} \tag{4.52}$$

$$\omega(k) = v_p(k)k \tag{4.53}$$

$$v_g(k) = \frac{d\omega}{dk} = v_p + \frac{dv_p}{dk}k \tag{4.54}$$

Here, ω and k are the angular frequency and the wave number, respectively. If the phase velocity is a decreasing function of k, the $\partial v_p/\partial k < 0$ term on the right-hand side of Eq. (4.54) makes $v_g < v_p$. This type of dispersion is referred to as normal dispersion. If the phase velocity is an increasing function of k, $v_g > v_p$. In this case, the dispersion is said to be anomalous. In the case of light, the phase velocity is expressed with the speed of light in a vacuum and the index of refraction, $v_p = c_0/n(k)$. We can conveniently discuss dispersion in association with the frequency dependence of the index of refraction.

4.3.2   Normal and Anomalous Dispersion

The index of refraction can be interpreted as the phase part of the interaction between light and a dielectric medium. See the paragraphs near Eq. (3.44). As Fig. 3.3 indicates, the refractive index increases with frequency from the red (low frequency) side toward the maximum value before the resonant frequency, sharply decreases toward the minimum value after the resonant frequency and starts increasing again toward the blue (high frequency) end of the spectrum. This tells us that in most frequency ranges, except for the range between the maximum and minimum near the resonance, the refractive index increases with frequency. In these frequency ranges, the dispersion is normal. In other words, the group velocity of the light is slower than the phase velocity. In the near resonance region where the refractive index has a negative slope to the frequency, however, the dispersion is anomalous and the group velocity is faster than the phase velocity. Below, we discuss this effect quantitatively.

Substituting the following expressions

$$v_p = \frac{c_0}{n} \tag{4.55}$$

$$\frac{dv_p}{dk} = c_0 \frac{d}{dk}\left(\frac{1}{n}\right) = c_0 \frac{d}{dn}\left(\frac{1}{n}\right)\frac{dn}{dk} = -\frac{c_0}{n^2}\frac{dn}{dk} \tag{4.56}$$

we can rewrite Eq. (4.54)in the following form:

$$v_g = v_p - \frac{c_0 k}{n^2}\frac{dn}{dk} \tag{4.57}$$

Equation (4.57) indicates that if $dn/dk > 0$, $v_g < v_p$ (normal dispersion), and if $dn/dk < 0$, $v_g > v_p$ (anomalous dispersion). We can say that in the spectral region on the red (low frequency) side of the resonance (where the refractive index is an increasing function of frequency), the dispersion is normal, and on the blue side near resonance where $dn/dk < 0$ it is anomalous. We can explicitly express the relationship between the type of dispersion and temporal frequency as below.

Using Eqs. (4.52), (4.54), and (4.55), we can put the second term on the right-hand side of Eq. (4.57) in the following form:

$$\frac{c_0 k}{n^2}\frac{dn}{dk} = \frac{c_0}{n^2}\frac{\omega}{v_p}\frac{dn}{d\omega}\frac{d\omega}{dk} = \frac{c_0}{n^2}\frac{\omega}{v_p}\frac{dn}{d\omega}v_g = \frac{c_0}{n^2}\frac{\omega n}{c_0}\frac{dn}{d\omega}v_g = \frac{\omega}{n}\frac{dn}{d\omega}v_g \tag{4.58}$$

From Eqs. (4.57) and (4.58), we obtain the following equation:

$$v_g = \frac{v_p}{1 + \frac{\omega}{n}\frac{dn}{d\omega}} \tag{4.59}$$

Equation (4.59) explicitly indicates that on the red side of the resonance where $dn/d\omega > 0$, the dispersion is normal as $v_g < v_p$, and on the blue side the dispersion is anomalous as $v_g < v_p$.

The above discussion indicates that in the near resonance range where the refractive index has a negative slope to the frequency, the group velocity is faster than the speed of light [33]. While it is true, in most dielectric media this frequency range overlaps with the high absorption frequency band, as seen in Fig. 3.3, and the anomalous dispersion is hardly observed. Also, notice that although this anomalous dispersion makes the group velocity higher than the speed of light, the phenomenon does not violate Einstein's view [34].

Probably the most obvious effect of dispersion is that an optical prism breaks white light into multiple colors. A less obvious and practically important effect is the fact that the refractive index depends on the frequency even in a range far away from the resonant frequency.[1] In an optical setup where high precision is required, we must be careful in selecting optical components whose performance depends on the frequency. For instance, when you use a calcite polarizer to remove unwanted polarization at a high extinction ratio, the frequency dependence of the refractive index can compromise the performance. It is important to design the polarizer by taking into account the optical frequency of the use.

4.4   Diffraction

Diffraction is known as the phenomenon that a wave goes around an obstacle. Near the edge of the obstacle, the wave changes its direction and propagates into the space behind the obstacle. Since the direction of the wave's propagation is perpendicular to the wavefront, this "going around" phenomenon can be interpreted as the wavefront is altered near the edge of the obstacle. When a wave passes through an opening, the wavefront is altered along the edge of the opening. If the size of the opening is comparable to the wavelength, the effect of wavefront alternation occurs in large parts of the cross-sectional area of the wave. Consequently, the wave's propagation characteristics are drastically changed by the aperture.

As a light wave, laser beams exhibit diffraction when they pass through an opening. When we use a laser beam in our experiment, often we need to reduce the beam diameter so that it passes through optical components that have small openings. If we use a diaphragm to reduce the beam diameter by clipping the outer area of the beam, it alters the propagation of the laser beam due to diffraction. Usually, this causes an undesired effect such as significant distortion of the wavefront or non-uniformity in the lateral intensity profile. We can use a focusing lens to reduce the beam size. However, the reduction in the beam radius also alters the beam divergence angle, as we discussed in Sect. 3.3.3. In addition, there is a theoretical limit on the reduction of the beam size (See Sect. 4.4.2).

4.4.1   Diffraction by a Single Slit

Diffraction is a combined effect of Huygens' principle [35–37] and interference. According to Huygens' principle, "every point on a wavefront is a source of the wave". The light waves from these multiple sources (called the secondary sources) interfere with each other and form an interference pattern called the diffraction pattern. Consider this mechanism using a simple case where a single slit diffracts a light wave.

Figure 4.16a shows that a light wave of wavelength λ passes through a slit of width d. The center of the slit is at the origin $(x, y, z) = (0, 0, 0)$, and a screen is placed perpendicular to the x-axis at $x = R$. Let θ be the angle formed by the x-axis and the line from the origin

---

[1] Normally, the frequency of the light used in a practical application is much lower than the resonant frequency of the optical component involved. So, it is safe to assume a positive slope $dn/d\omega$ in most applications.

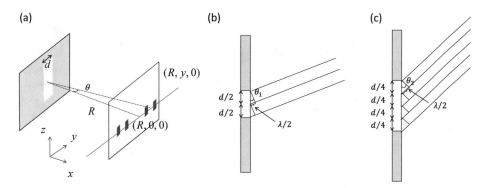

Fig. 4.16 Diffraction by a single slit **a** Single slit and screen; **b** Condition for first destructive interference; **c** Condition for second destructive interference. Slit width is enlarged for better visibility

to point $(R, y, 0)$ on the screen. According to Huygens' principle, we consider destructive interference on the screen as a function of θ. When $\theta = 0$, it is obvious that all the secondary sources (the element waves on the slit) inside the slit are in phase. No destructive interference occurs on the screen. As we increase θ in the positive y-direction, the distance to the screen from each secondary source increases. The greatest difference is between the secondary source closest to the positive y side of the slit edge and the negative y side of the slit edge. As Fig. 4.16b illustrates, the first destructive interference occurs when θ is increased to a specific value θ_1 at which the distance from the secondary source on the negative y side of the slit is greater than that from the positive y side of the slit by half wavelength $\lambda/2$. This condition can be expressed as follows:

$$\frac{d}{2} \sin \theta_1 = \frac{\lambda}{2} \tag{4.60}$$

As you can easily imagine, the next destructive interference occurs when the angle θ is further increased to θ_2. As indicated by Fig. 4.16c, this condition involves four secondary sources in the slit. In this fashion, we can express mth destructive interference by the following expression:

$$\frac{d}{2m} \sin \theta_m = \frac{\lambda}{2} \tag{4.61}$$

Further analysis yields the following expression for the intensity profile of the diffraction pattern on the screen. This type of diffraction is known as Fraunhofer diffraction [38, 39].

$$I(\theta) = I(0) \left(\frac{\sin \beta}{\beta} \right)^2 \tag{4.62}$$

Here, β is a function of the ratio of the slit width to the wavelength and angle θ.

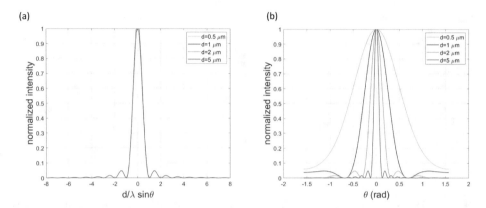

Fig. 4.17 Diffraction by a single slit **a** Intensity profile of single slit diffraction as a function of β/π; **b** Intensity profile of single slit diffraction as a function of θ

$$\beta = \frac{kd}{2}\sin\theta = \frac{\pi d}{\lambda}\sin\theta \tag{4.63}$$

Figure 4.17a plots the intensity as a function of $\beta/\pi = (d/\lambda)\sin\theta$ for various slit widths for a wavelength of $\lambda = 633$ nm (He–Ne laser's wavelength). In this format, all the intensity profiles resulting from different slit widths lie on the same curve. This independence of the slit width comes from the fact that the parameter β contains d. Figure 4.17b plots the same intensity profile as a function of angle θ. It illustrates the fact that the central lobe of the intensity profile increases as the slit width decreases. Note that the broadest intensity profile is the case where the wavelength (633 nm) is longer than the slit width (500 nm). The other profiles are when the wavelength is shorter than the slit width.

4.4.2   Diffraction by a Circular Aperture and Diffraction Limit

The intensity pattern of diffraction due to a circular aperture is given as follows:

$$I(\theta) = I(0)\left(2\frac{j_1(\beta)}{\beta}\right)^2 \tag{4.64}$$

In Eq. (4.64), $j_1(\beta)$ is the first-order Bessel function of the first kind and β is defined by Eq. (4.63). From the property of the Bessel function of this kind, the first zero occurs when $\beta = 3.8317$ indicating that the following condition defines the central lobe in this case:

$$\Delta\theta \approx \sin\theta_{min} = 3.8317\frac{\lambda}{\pi d} = 1.22\frac{\lambda}{d} \tag{4.65}$$

Similar to the single slit case, the size of the central lobe of the diffraction pattern by a circular aperture is determined by the wavelength relative to the slit size. The central lobe is defined by the first zero corresponding to the minimum angle θ_{min}.

This minimum angle is associated with an important concept known as the diffraction limit. Call the radius of the central lobe (distance from the center of the central lobe to the first zero) a and the distance from the aperture to the screen R. From Fig. 4.16, we find that $a = R \tan \theta_{min} \cong R \sin \theta_{min}$. Thus, using Eq. (4.65), we can express the radius of the central robe as follows:

$$a = 1.22 \frac{R\lambda}{d} \tag{4.66}$$

Equation (4.66) indicates that light of wavelength λ going through an aperture with diameter d has a minimum radius a at the screen R (m) away. Here d can be interpreted as the size of the light source and R be the distance from the source to the image. In the context of focusing a laser beam with a positive lens of diameter D and focal length f, we can use $R = f, d = D$. Thus, we obtain the following equation for the minimum diameter $2a$ of the laser beam at the focus.

$$2a = 2 \times 1.22 \frac{f\lambda}{D} = 2.44 F\lambda \tag{4.67}$$

Here, $F = f/D$ is called the F number of the lens. Equation (4.67) indicates that the laser beam of wavelength λ cannot be focused into a smaller radius than a with a positive lens of F number, $F = f/D$.

4.4.3 Diffraction of a Laser Beam

Low beam divergence is one of the important properties of lasers. However, the output beam from any laser system diverges due to diffraction. As we discuss in the next chapter, normally a low divergence laser beam is a Gaussian beam of the lowest mode known as TEM_{00} mode. This mode is characterized by the minimum beam radius known as the beam waist size. For a given beam waist size, the diverging property obeys certain characteristics. In other words, the diffraction is determined once the beam waist is given. We can view the beam waist as corresponding to the slit size d in the above argument. In fact, the beam divergence angle of a TEM_{00} mode is inversely proportional to the beam waist size w_0 [40].

$$\theta_{beam} = \tan^{-1}\left(\frac{\lambda}{\pi w_0}\right) \cong \frac{\lambda}{\pi w_0} \tag{4.68}$$

The above fact that a given Gaussian beam obeys this diffraction condition means that we cannot make the laser beam from a given laser system by passing it through an aperture. As we discussed above, once a laser beam passes through an aperture, diffraction occurs. After passing through the aperture, the laser beam forms a diffraction pattern and the optical energy is no longer confined to a small diameter as a single beam. If we need to change the diameter of a Gaussian beam, it is necessary to convert it to another Gaussian beam that has a smaller beam waist size. However, we need to be careful because a smaller beam waist causes the beam to diverge faster as it propagates through the space (Eq. (4.68)). For further details about Gaussian beams and their diffraction property, see the next section.

References

1. E. Hecht, *Optics* 4th edn. (Addison Wesley, San Francisco, CA, USA, 2002) pp. 100–104.
2. 1.3 Refraction, openstax, https://openstax.org/books/university-physics-volume-3/pages/1-3-refraction
3. E. Hecht, *Optics* 4th edn. (Addison Wesley, San Francisco, CA, USA, 2002) pp. 116.
4. D. J. Griffiths, *Introduction to electrodynamics*. 3rd edn. (Prentice Hall, Upper Saddle River, NJ, USA, 1999) pp. 331–333
5. 6.7 Stokes' Theorem, openstax, https://openstax.org/books/calculus-volume-3/pages/6-7-stokes-theorem
6. Polarization Control with Optics, https://www.newport.com/n/polarization-control-with-optics#:%CB%9C:text=S%2Dpolarization%20refers%20to%20the,remaining%20figures%20of%20the%20section (accessed on August 7, 2022)
7. M. S. Kao, C. F Chang, *Understanding Electromagnetic Waves* 1st edn. (Springer Nature, Cham, Switzerland, 2020) pp. 219
8. D. J. Griffiths, *Introduction to electrodynamics*. 3rd edn. (Prentice Hall, Upper Saddle River, NJ, USA, 1999) Sec. 9.5, pp. 407–411
9. V. Rojansky, *Electromagnetic fields and waves* (Dover Pub, Inc., New York, 1979), Chap. 26, pp. 410–428
10. J. C. Slater, N. H. Frank, *Electromagnetism*, (Dover Pub, Inc., New York, 1969) Chap. XI, pp. 129–146
11. S. Yoshida, *Waves; Fundamental and dynamics* (Morgan & Claypool, San Rafael, CA, USA, IOP Publishing, Bristol, UK, 2017), p.3–7
12. Engineering Acoustics/Reflection and transmission of planar waves https://en.wikibooks.org/wiki/Engineering_Acoustics/Reflection_and_transmission_of_planar_waves (accessed on August 4, 2022)
13. E. Hecht, *Optics* 4th edn. (Addison Wesley, San Francisco, CA, USA, 2002) pp. 113–122
14. E. Hecht, Amplitude Transmission Coefficient for Internal Reflection. Am. J., Phys. **41**, 1008-1010, 1973.
15. CVI Optical Components and Assemblies, http://122.1.221.126/product/optics/documents/CVI_2006_catalog.pdf (accessed on August 10, 2022)
16. D. F. Vanderwerf, *Applied Prismatic and Reflective Optics* pp. 63-64, https://doi.org/10.1117/3.867634.ch3   https://www.spiedigitallibrary.org/ebooks/PM/Applied-Prismatic-and-Reflective-Optics/Chapter3/Polarization-Properties-of-Prisms-and-Reflectors/10.1117/3.867634.ch3 (accessed on August 10, 2022)
17. E. Hecht, *Optics* 4th edn. (Addison Wesley, San Francisco, CA, USA, 2002) Chap 8, pp. 325–384
18. G. Chartier, *Introduction to Optics* (Springer, New York, 2005) pp 179–258
19. Y. Otani, 26 Birefringence Measurement, T. Yoshizawa, Ed. *Handbook of Optical Metrology*, 2nd, edn. O'Reilly https://www.oreilly.com/library/view/handbook-of-optical/9781466573598/xhtml/37_Chapter26.xhtml (accessed on August 7, 2022)

20. Calcite, GEOLOGYSCIENCE, https://geologyscience.com/minerals/calcite/ (accessed on August 10, 2022)
21. S. Yoshida, S. Yoshida, *Waves; Fundamental and dynamics* (Morgan & Claypool, San Rafael, CA, USA, IOP Publishing, Bristol, UK, 2017) pp. 2–28 – 2–29
22. A. Labeyrie, S. G. Lipson, P. Nisenson, *An Introduction to Optical Steller Interferometry* (Cambridge Univ. Press, Cambridge, UK, 2006) pp. 1–63
23. D. F. Buscher, *Practical Optical Interferometry, Imaging at Visible and Infrared Wavelengths* (Cambridge Univ. Press, Cambridge, UK, 2015)
24. M. Lucki, L. Bohac, R. Zeleny, *Fiber Optic and Free Space Michelson Interferometer — Principle and Practice*, InTech Open, https://doi.org/10.5772/57149,2014a.
25. M. Lucki, L. Bohac, R. Zeleny, *Fiber Optic and Free Space Michelson Interferometer — Principle and Practice*, InTech Open, https://doi.org/10.5772/57149,2014b.
26. S. Yoshida, S. Adhikari, K. Gomi, R. Shrestha, D. Huggett, C. Miyasaka, I. Park, Opto-acoustic technique to evaluate adhesion strength of thin-film systems, AIP Advances **2**, 022126-1 - 022126-7, 2012
27. E. Hecht, *Optics* 4th edn. (Addison Wesley, San Francisco, CA, USA, 2002) Chapter 8, pp. 325–384.
28. W. Lauterborn, T. Kurz, *Coherent Optics* 2nd edn., (Springer, Berlin, New York, 1993)
29. J. C. Dainty, Ed. *Laser Speckle and Related Phenomena* (Springer, Berlin, Heidelberg, 1975)
30. C. A. Sciammarella, F. M. Sciammarella, Experimental Mechanics of Solids. (Wiley, Hoboken, NJ, USA) 2012.
31. J. A. Leendertz, Interferometric displacement measurement on scattering surface utilizing speckle effect. J. Phys. E **3**, 214–218, 1970
32. P. Meinlschmidt, K. D. Hinsch, R. S. Sirohi: Selected Papers on Electronic Speckle Pattern Interferometry: Principles and Practice. Milestone Series, vol. 132, SPIE Press, Bellingham, Washington, USA, 1996.
33. L. J. Wang, A. Kuzmich, A. Dogariu, Gain-assisted superluminal light propagation, Nature **406** 277–279, 2000
34. A. Einstein, *The meaning of Relativity*, 5th edn., (MJF Books, New York, 1984)
35. J. C. Slater, N. H. Frank, *Electromagnetism*, (Dover Pub, Inc., New York, 1969) Chap. XIII, pp. 167–178
36. E. Hecht, *Optics* 4th edn. (Addison Wesley, San Francisco, CA, USA, 2002) pp. 104–105
37. D. A. B. Miller, Huygens's wave propagation principle corrected, Op. Lett., **16** (18), 1370–1372, 1991
38. E. Hecht, *Optics* 4th edn. (Addison Wesley, San Francisco, CA, USA, 2002) pp. 452–457
39. J. C. Slater, N. H. Frank, *Electromagnetism*, (Dover Pub, Inc., New York, 1969), Chap XIV, pp. 180–192
40. R. Pachotta, Beam Divergence, RP Photonics Encyclopedia, https://www.rp-photonics.com/beam_divergence.html (accessed on August 10, 2022)

Lasers

5

5.1 Laser in a Nutshell

The word laser stands for Light Amplification of Stimulated Emission of Radiation. Stimulated emission [1] is a process in which a photon interacts with a higher energy state of an atomic (or molecular) system and releases another photon of the same energy; one photon results in two photons. Thus, there is a net gain of one photon per stimulated emission. This gain is the source of the "amplification" of radiation.

When a lasing species has an intrinsically high gain, stimulated emission alone amplifies the optical power to a considerable degree. (We call this type of amplification Amplified Spontaneous Emission, ASE.). However, ASE does not necessarily exhibit high coherency. [2] By placing a gain medium inside an optical resonator, we can obtain a higher gain and coherency in a controlled fashion.

Thus, laser physics has two main elements; atomic (molecular) physics and optical resonator dynamics. We discuss these elements separately in the following sections.

5.2 Atomic Physics

5.2.1 Atomic Systems as a Resonator

According to quantum mechanics [3, 4], the wave function describes the behavior of a microscopic particle like an electron. A wave function is a solution to the Shrödinger equation and is a complex number. The square of its magnitude represents the probability density. We can find physical quantities, such as the displacement from the equilibrium position and momentum, as the expectation value using the probability density function.

© The Author(s), under exclusive license to Springer Nature Switzerland AG 2023   135
S. Yoshida, *Fundamentals of Optical Waves and Lasers*, Synthesis Lectures on Wave
Phenomena in the Physical Sciences. https://doi.org/10.1007/978-3-031-18188-7_5

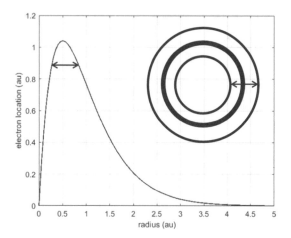

Fig. 5.1 Expected value of electron orbiting nucleus as a function of radius

This concept of probability function comes from Heisenberg's uncertainty principle [5, 6]. According to this principle, we cannot identify the position and velocity of a microscopic particle at the same time. This situation is contrastive to classical objects like a billiard ball. If a billiard ball moves on the inner surface of a spherical shell, it keeps its distance from the center of the sphere. Microscopic particles do not behave like this billiard ball. Consider the electron orbiting around a proton. Although its average distance from the proton is constant, the electron deviates from this distance as it orbits. Consequently, the electron's trajectory looks like a cloud. We call this cloud the electron cloud. Figure 5.1 illustrates the expected value of the radius of electron orbiting a nucleus two-dimensionally. The insert is a schematic illustration to indicate the electron cloud has a higher density near the radius corresponding to the peak expected value.

This difference between the electron and billiard ball comes from the situation where the probability density is a wave function. The behavior of the electron around the average radius resembles the harmonic oscillation of a spring–mass system. The electron is always pulled back to the average radius when it deviates to its longer or shorter side. One difference from the harmonic oscillation case is that the deviation from the equilibrium is not symmetric as Fig. 5.1 indicates. This difference is due to the asymmetry of the Coulomb force as a function of the radius. The inverse proportionality to the radius makes the Coulomb force stronger when the electron deviates toward the nucleus than away from it. In contrast, the magnitude of the spring force is the same in both directions from the equilibrium.

Another difference is that the electron's motion is a wave phenomenon, as discussed above. From this viewpoint, we can say that the electron's behavior around the average radius is analogous to the standing wave presented in Fig. 1.11a, where a standing wave forms under a fixed-end condition. The boundary condition confines the wave function in a range of radii. When an electron is placed in a symmetric, infinite potential well [7], the wave function takes the identical form to the standing wave solution of Fig. 1.11 situation.

Recall that in Fig. 1.11a and b the standing waves form under resonant conditions where the distance between the fixed ends is equal to an integer multiple of the half wavelength. Since the wave velocity is constant, reducing the wavelength in half doubles the frequency. In this fashion, the resonant frequency varies with the mode number N as $f^N = Nf_0$. Here, f_0 is the increment of the frequency series.

This interpretation explains qualitatively why the electronic state changes discretely with a constant increment in frequency. It also illustrates the resonant energy exchange between an atomic system and light waves under resonance. When an atom absorbs energy from the incident light, we call the phenomenon photon absorption. When the direction of energy transfer is opposite, we call it photon emission. We will discuss more photon absorption and emission in the next section.

5.2.2   Absorption and Emission

The atom–electromagnetic field interaction is a resonant process. Earlier in this book, we discussed that an electron orbiting a nucleus absorbs electromagnetic energy from an incident light wave and changes its orbit to a more energetic one. (See "Electric dipole and interaction with light" and Fig. 2.3). We call the original and the resultant states the lower and the upper states, respectively. (In the case the atom is initially at the ground state, we may call the upper state the excited state.) Here, the absorbed energy is equal to the energy difference between the lower and upper states. In the reverse process, the electron in the upper state returns to the lower state giving the corresponding energy back to the electromagnetic field. This process is called emission. In this fashion, the atomic system and the electromagnetic field exchange energy. As resonant processes, the associated absorption or emission process exhibits the same type of bell-shaped spectrum in the frequency domain as the mass–spring system example discussed in Chap. 1. That is why the frequency dependence of the energy absorption by a dipole (Fig. 3.3) looks like that of a spring–mass system driven by a sinusoidal force (Fig. 1.4). Figure 5.2 illustrates three possible forms of transitions between the lower and upper states. Figure 5.2 schematically illustrates three possible forms of transitions between the lower and upper states.

As indicated in Fig. 5.2, the atomic–electromagnetic interaction has two types of emission. The first is the spontaneous emission, and the second is the stimulated emission. The spontaneous emission occurs spontaneously when the radiative lifetime of the upper state elapses. Being a random process, this form of emission does not contribute to the amplification or coherence of the emission.

In the process of stimulated emission, an atom in the upper state goes back to the lower state through an interaction with a photon whose energy is equal to the energy difference between the upper and lower states. This process may sound counterintuitive; why doesn't the upper state absorb the photon and go up to a higher energy state instead of down to the

Fig. 5.2 a Stimulated
emission; **b** Spontaneous
emission; **c** Absorption

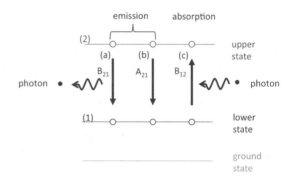

lower state? We need to use quantum mechanics to answer this question precisely. However, the following qualitative explanation will help us grasp the gist of the mechanism.

Consider that an upper state atom is steadily oscillating at frequency ω_2 when a photon of frequency Ω is incident. Note that this atom was originally at the lower state with frequency ω_1. Therefore, $\omega_2 = \omega_1 + \Omega$. For this steadily oscillating atom, the electromagnetic field carried by the photon is a disturbance. It tends to lose the stable energy status. It could absorb the photon energy to get further excited if $\omega_2 + \Omega$ is an allowed state. If $\omega_2 + \Omega$ is not an allowed state, the disturbance can push this atom back to the lower state by having it lose the energy corresponding to Ω. Here, since $\omega_2 - \Omega = \omega_1$ the destination is the initial state, i.e., it is an allowed state.

A simple mathematical identity $\cos\alpha\cos\beta = (1/2)(\cos(\alpha + \beta) + \cos(\alpha - \beta))$ may help understand the above mechanism. We can represent the upper state oscillation with $\cos\omega_1 t$. The incident photon disturbs this oscillation with $\cos\Omega t$. The mathematical identity indicates that this disturbance alters the frequency from ω_1 to $\omega_1 \pm \Omega$. Of these two frequencies, $\omega_1 - \Omega = \omega_0$ is allowed. By the way, if $\omega_1 + \Omega$ is an allowed state, the lower state atom can absorb two photons of Ω simultaneously. We call this process a multi-photon (in this case, two-photon) absorption [8].

Photon energy and uncertainty principle

In the above paragraphs, we used the word photon several times without defining it. Here, we take a moment to consider the concept of the photon. There are multiple ways to describe a photon, such as a particle feature of light, a quantum particle with no mass, the quantum of the electromagnetic field, etc. In my opinion, we can define the photon through interaction with matters. As discussed above, atoms can transfer energy to and from an electromagnetic field through resonant processes. The transferred energy is quantized, and the increment of quantized energy is associated with the lowest frequency (called f_0 in the above analogy to harmonic oscillation). This observation leads to the interpretation that the frequency of the photon energy to the lowest excitation state from the ground state (ν of $h\nu$) as corresponding to f_0.

This discussion makes us wonder about the physical meaning of the Planck constant h in conjunction with the photon energy and harmonic oscillation. Heisenberg's uncertainty principle [5, 6] states that the uncertainty in the product of position and velocity of microscopic particles is of the order of h. This situation indicates that somehow we should be able to explain the uncertainty principle using the spring–mass oscillation. In my opinion, the following explanation is legitimate. Consider when a mass connected to a spring is stationary when it is under harmonic oscillation. The answer is at the two turning points. At these points, the displacement from the equilibrium is either at the rightward or leftward maximum, i.e., the uncertainty is the greatest (from the negative to positive extremum). On the other hand, when the displacement is null at the equilibrium, the velocity is either leftward or rightward maximum, i.e., its uncertainty is the highest.

5.2.3 Optical Energy of Induced Transition

In the preceding section, we discussed the mechanism of absorption and emission processes. Here, we consider the optical energy associated with these transition processes.[1] First, consider the induced processes where a photon is involved. The probability (the occurrence per unit time) of the stimulated emission and absorption is proportional to the spectral optical energy density $\rho(\nu)$.

$$(W'_{21})_{induced} = B_{21}\rho(\nu) \tag{5.1}$$

$$(W'_{12})_{induced} = B_{12}\rho(\nu) \tag{5.2}$$

Here, $(W'_{21})_{induced}$ and $(W'_{12})_{induced}$ are the probability of transition due to induced emission and absorption, respectively, and ν is the optical frequency of the incident light. The constant of proportionality, $B_{21} = B_{12}$ is called Einstein's B coefficient.

The total transition from State 2 to 1 and State 1 to 2 become as follows:

$$(W'_{21}) = B_{21}\rho(\nu) + A_{21} \tag{5.3}$$

$$(W'_{12}) = (W'_{21})_{induced} = B_{21}\rho(\nu) \tag{5.4}$$

When an atomic system is in thermal equilibrium with the field of blackbody radiation [9] at temperature T, we can express the power density as follows (See Chap. 5 of [10]):

$$\rho(\nu) = \frac{8\pi h\nu^3}{c^3} \frac{1}{exp[h\nu/kT] - 1} \tag{5.5}$$

The probability of transition from 2 to 1 must be the same as 1 to 2. (See Chap. 5 of [10])

$$N_2 W'_{21} = N_1 W'_{12} \tag{5.6}$$

---

[1] See Chap. 5 of [10] for more discussions of this section.

Putting Eq. (5.5) in Eqs. (5.3) and (5.4) and using Eq. (5.6), we obtain the following equality:

$$N_2 \left[B_{21} \frac{8\pi h v^3}{c^3 (e^{hv/(kT)} - 1)} + A_{21} \right] = N_1 \left[B_{12} \frac{8\pi h v^3}{c^3 (e^{hv/(kT)} - 1)} \right] \quad (5.7)$$

Because the atomic system is in thermal equilibrium, the number density obeys the Boltzman distribution [11].

$$\frac{N_2}{N_1} = e^{-hv/(kT)} \quad (5.8)$$

From Eqs. (5.7) and (5.8), we obtain the following expression:

$$\frac{8\pi h v^3}{c^3 (e^{hv/(kT)} - 1)} = \frac{A_{21}}{B_{12} e^{hv/(kT)} - B_{21}} \quad (5.9)$$

Equation (5.9) holds when the following two conditions are true:

$$B_{12} = B_{21} \quad (5.10)$$

$$\frac{A_{21}}{B_{21}} = \frac{8\pi h v^3}{c^3} \quad (5.11)$$

From Eq. (5.10), we can set $(W'_{12})_{induced} = (W'_{21})_{induced} \equiv W'_i$. With this expression and Eq. (5.11), we can rewrite Eq. (5.1) as follows:

$$W'_i = \frac{A_{21} c^3}{8\pi h v^3} \rho(v) = \frac{c^3}{8\pi h v^3 t_{spont}} \rho(v) \quad (5.12)$$

Here, $t_{spont} = 1/A_{21}$ denotes the spntaneous emission lifetime.

By expressing the electromagnetic energy density that induces the transition from State 2 to 1 with ρ_v and the spectral density function of the electromagnetic energy of the incident field with $g(v)$, we can express Eq. (5.12) as follows:

$$W'_i = \frac{c^3 \rho_v}{8\pi h v^3 t_{spont}} g(v) \quad (5.13)$$

Here,

$$\rho(v) = \rho_v g(v) \quad (5.14)$$

In Eq. (5.14), $\rho(v)$ is the energy density per frequency and, therefore, its unit is J/m^3/Hz = J $\cdot$ s/m^3. The function $g(v)$ is the spectral density function in 1/Hz = s. So, the unit of ρ_v is J/m^3; it represents the optical energy density.

This observation leads to the following expression for the intensity of the light wave (laser beam) associated with the emission probability W'_i. Letting c be the speed of light, we can express the optical intensity as the product of c and the energy density ρ_v.

$$I = c\rho_v \quad (5.15)$$

Substitution of Eq. (5.15) into Eq. (5.13) yields the following expression of the transition probability W_i' with intensity I:

$$W_i' = \frac{c^2 I}{8\pi h \nu^3 t_{spont}} g(\nu) \tag{5.16}$$

Equation (5.16) indicates that the probability of the induced transition is proportional to the light intensity and the spectral density function (i.e., the shape of the spectrum) and inversely proportional to the spontaneous emission lifetime. The shorter the radiative lifetime of the upper state, the higher the optical intensity, and the sharper the spectral density function at the transition frequency, the higher the transition probability.

Spectral shape of spontaneous emission

It is worthwhile considering the spectral shape of spontaneous emission when the broadening is due to a finite lifetime of the upper state. In short, the spectral shape is the so-called Lorentzian. Here, we discuss why it is a Lorentzian.

Let τ be the upper state lifetime, and $\sigma/2 = \tau^{-1}$ be the associated decay coefficient. As discussed in Sect. 1.1.2, we can express the electric field of the emission as follows:

$$e(t) = A_0 e^{-i\frac{\sigma}{2}t} \cos \omega_0 t = \frac{A_0}{2} \left\{ e^{i(\omega_0 + i\frac{\sigma}{2})t} + e^{-i(\omega_0 - i\frac{\sigma}{2})t} \right\} \tag{5.17}$$

Here, ω_0 is the resonant frequency of the emission. The Fourier transform of $e(t)$ is

$$E(\omega) = \int_{-\infty}^{\infty} e(t) e^{-i\omega t} dt = \int_0^{\infty} \frac{A_0}{2} \left\{ e^{i(\omega_0 + i\frac{\sigma}{2})t} + e^{-i(\omega_0 - i\frac{\sigma}{2})t} \right\} e^{-i\omega t} dt$$

$$= \frac{A_0}{2} \left\{ \frac{1}{i(\omega - \omega_0) - \frac{\sigma}{2}} + \frac{1}{i(\omega + \omega_0) - \frac{\sigma}{2}} \right\}. \tag{5.18}$$

The spectral intensity of the emission is proportional to the square of the Fourier transform. Since the second term of the right-hand side of Eq. (5.18) is significantly smaller than the first term, we can consider only the first term.

$$|E(\omega)|^2 \propto \frac{1}{(\omega - \omega_0)^2 + (\sigma/2)^2} \tag{5.19}$$

Normalizing the spectral density function as

$$\int_{-\infty}^{\infty} g(\nu) d\nu = 1, \tag{5.20}$$

and using linewidth $\Delta\nu = \sigma/(2\pi)$, we obtain the following expression for $g(\nu)$:

$$g(\nu) = \frac{\Delta\nu}{2\pi \left[(\nu - \nu_0)^2 + (\Delta\nu/2)^2 \right]} \tag{5.21}$$

Here, $\Delta\nu$ is the FWHM (Full Width at Half Maximum) of the spectral density function. We will revisit this topic later under the "Active resonator" section when we discuss the gain spectral width.

5.2.4 Optical Gain

Next, consider the optical gain as the light beam propagates through a gain medium. The increase in the intensity of a laser beam propagating in the positive z direction, dI, over distance dz is equal to the total optical power in the volume made by a unit cross-sectional area (because the intensity is the power per unit cross-sectional area) and length dz. Here the optical power is the product of the net induced emission (the induced emission minus induced absorption), the transition probability, and the photon energy $h\nu$.

$$dI = (N_2 - N_1)W_i'(h\nu)dz \qquad (5.22)$$

Substituting Eq. (5.16) into W_i', we can write Eq. (5.22) in the differential form:

$$\frac{dI}{dz} = (N_2 - N_1)\frac{c^2 I}{8\pi\nu^2 t_{spont}}g(\nu) \qquad (5.23)$$

We can view Eq. (5.23) as a differential equation of the following form:

$$\frac{dI(z)}{dz} = \gamma I(z) \qquad (5.24)$$

where γ is a constant.

$$\gamma = (N_2 - N_1)\frac{c^2}{8\pi\nu^2 t_{spont}}g(\nu) \qquad (5.25)$$

We can easily solve the differential equation (5.24) as follows:

$$I(z) = I(0)e^{\gamma z} \qquad (5.26)$$

From Eq. (5.25), we find the following condition for the sign of γ:

$$\gamma > 0 \quad \text{if} \quad N_2 > N_1 \qquad (5.27)$$
$$\gamma < 0 \quad \text{if} \quad N_2 < N_1 \qquad (5.28)$$

Conditions (5.27) and (5.28) tell us that when the upper state is more populated than the lower state, the gain is positive; otherwise, it is negative. The condition where an atomic system has a higher population in the upper state is called population inversion.

Fig. 5.3 Three-level system. (1)–(3) indicate State 1–State 3

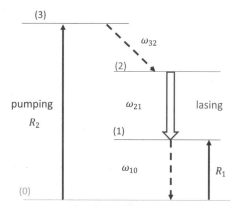

5.2.5 Population Inversion and Rate Equation

For an atomic or molecular system to act as a lasing source, it must have positive optical gain, i.e., it must establish population inversion. In thermal equilibrium, the number of atoms with higher energy is always lower than the lower energy. Thus, population inversion cannot occur.

To establish population inversion, we need to provide the atomic system with energy from an external source. We say that we pump the atomic system to obtain positive gain via population inversion. Any form of energy can be used as a pumping source as far as it can invert the population. Depending on the form of energy, we call the laser system optically pumped, chemically pumped, gas-dynamically pumped lasers, etc.

Suppose that the pumping mechanism excites atoms in the ground states to State 3 in Fig. 5.3. By a non-radiative process, the atoms decay to State 2, the upper lasing state. Atoms in the upper state emit light and go to the lower lasing state (State 1). Atoms in State 1 decay to the ground state. To establish population inversion, the population in State 2 must be greater than that in State 1. To achieve the situation, the pumping rate R_2, decay rates ω_{32} and ω_{10} should be high, and the decay rate ω_{21}, which competes with the lasing action, should be low.

The population densities at State (2) and State (1) obey the following rate equations:

$$\frac{dN_2}{dt} = R_2 - N_2\omega_{21} - W_i(N_2 - N_1) \tag{5.29}$$

$$\frac{dN_1}{dt} = R_1 - N_1\omega_{10} + W_i(N_2 - N_1) + N_2\omega_{21} \tag{5.30}$$

In a steady state, the following condition holds:

$$\frac{dN_2}{dt} = \frac{dN_1}{dt} = 0 \tag{5.31}$$

Under condition (5.31), we can solve Eqs. (5.29) and (5.30) to find N_1 and N_2.

$$N_1 = \frac{R_1 + R_2}{\omega_{10}} \tag{5.32}$$

$$N_2 = \frac{\{W_i(R_1 + R_2)/\omega_{10}\} + R_2}{W_i + \omega_{21}} \tag{5.33}$$

So,

$$N_2 - N_1 = \frac{R_2\{1 - (\omega_{21}/\omega_{10})(1 + R_1/R_2)\}}{W_i + \omega_{21}} \equiv \frac{R}{W_i + \omega_{21}} \tag{5.34}$$

Equation (5.34) indicates that the greater the pumping rate to the upper state R_2, the decay rate of the lower state ω_{10} and the lower the pumping rate to the lower state R_1 and the decay rate of the upper state ω_{21}, the population inversion $N_2 - N_1$ becomes higher. We call the quantity R defined on the right-hand side of Eq. (5.34) the pumping rate. For laser species that have short radiative lifetimes such as excimer lasers [12], the decay rate from the upper state ω_{21} is high. These laser species, therefore, require a stronger pumping rate R to establish the same level of population inversion.

Above, we discussed the gist of the mechanism to generate population inversion using a three-level system as an example. For more detailed discussions of this section, I encourage the reader to consult references, e.g., Chap. 6 of [10].

5.3   Optical Resonator

As clear from the discussion in the earlier sections of this chapter, the mechanism of stimulated emission is the source of amplification and coherence. When stimulated emission occurs in a gain medium, the resultant photons stimulate upper state atoms. This process repeats as long as the upper state has more population than the lower state, producing new photons of the same frequency. However, the amplification and coherence generated by stimulated emission alone are insufficient for practical applications. It is where an optical resonator plays an important role. By placing the gain medium in a resonator, we can enhance the amplification and coherence.

In this section, we discuss how optical resonators operate. We start the discussion with a one-dimensional passive resonator to explore the resonant condition. Next, we extend the discussion to a two-dimensional passive resonator to consider the transverse mode of resonance. Finally, we place a gain medium in a two-dimensional configuration and discuss how a laser resonator produces optical power that we can use.

Fig. 5.4 Fabry-Perot etalon

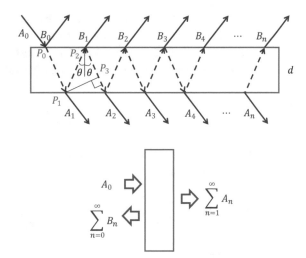

5.3.1 Fabry-Perot Etalon

Figure 5.4 illustrates the concept of a one-dimensional optical resonator known as Fabry–Perot etalon [13] schematically (See also Chap. 4 of [10]). Here, the top illustration is the case when the incident beam is oblique to the surface of a slab of glass. Part of the incident beam is reflected on the front surface and the rest is transmitted through the slab and reaches the rear surface. At the rear surface, part of the beam is reflected and the rest is transmitted through to air. Consider the optical path difference between transmitted beams A_1 and A_2. The former travels inside the etalon from point P_0 to P_1. The latter travels for the extra distance P_1 through P_2 to P_3. Hence, the path difference is as follows:

$$\delta l = \frac{d}{\cos\theta} + \frac{d}{\cos\theta}\cos 2\theta = 2d\cos\theta \tag{5.35}$$

Here, d is the width of the etalon and θ is the angle of reflection. Thus, the phase difference δ is

$$\delta = 2\pi\frac{\delta l}{\lambda} = \frac{4\pi d\cos\theta}{\lambda} \tag{5.36}$$

The bottom illustration in Fig. 5.4 represents the case when the angle of incidence is zero. In this case, the phase difference per round trip between the two surfaces of the slab becomes $4\pi d/\lambda$. If the etalon width d is an integer multiple of half of the wavelength $\lambda/2$, the resonant condition discussed in Fig. 1.11a holds, indicating that the etalon acts as an optical resonator.

Denoting the round trip phase difference with δ, we obtain the following expressions for reflection:

$$B_0 = rA_0, \quad B_1 = tt'r'e^{i\delta}A_0, \quad B_2 = tt'r'^3e^{i2\delta}A_0, \ldots, B_n = tt'r'^{(2n-1)}e^{in\delta}A_0, \ldots \tag{5.37}$$

Here, r and t are the reflection and transmission coefficients at the front surface (air to glass), and r' and t' are those at the rear surface (glass to air).

Since the light ray reduces the amplitude by a factor rr' and gains a phase $e^{i\delta}$ every time it makes a round trip inside the etalon, we obtain the following recursive expression for B_i:

$$B_0 = rA_0 \tag{5.38}$$

$$B_n = r'^2 e^{i\delta} B_{n-1} \quad (n \geq 2) \tag{5.39}$$

We can express the amplitude of the total reflection, A_r, as the addition of the first reflection B_0 and the sum of the geometric progression for the other reflections.

$$A_r = rA_0 + \sum_{n=1}^{\infty} \left(r'^2 e^{i\delta} B_n \right) = rA_0 + \lim_{n\to\infty} B_n \frac{1 - (r'^2 e^{i\delta})^n}{1 - r'^2 e^{i\delta}}$$

$$= \left\{ r + tt'r' \frac{e^{i\delta}}{1 - r'^2 e^{i\delta}} \right\} A_0 = r \frac{1 - e^{i\delta}}{1 - r^2 e^{i\delta}} A_0 = \frac{(1 - e^{i\delta})\sqrt{R}}{1 - Re^{i\delta}} A_0 \tag{5.40}$$

Here, in going to the second line of Eq. (5.40), we used Eq. (5.37) for $n = 1$, and $\lim_{n\to\infty}(r'^2 e^{i\delta})^n = 0$ as $0 < r'^2 < 1$. In the second line, we used $r = -r'$, $r^2 + tt' = 1$, and $R = r^2$.

Similarly, for the transmitted light, we find $A_1 = tt'A_0$ and $A_{n+1} = rr'e^{i\delta} A_n$, and obtain the following expression.

$$A_t = \lim_{n\to\infty} tt'A_0 \frac{1 - (rr'e^{i\delta})^n}{1 - rr'e^{i\delta}} = tt'A_0 \frac{1}{1 - rr'e^{i\delta}} = A_0 \frac{T}{1 - Re^{i\delta}} \tag{5.41}$$

Here, $tt' = T$.

Remembering that the optical intensity is the product of the amplitude and its complex conjugate, and noting $(1 - e^{i\delta})(1 - e^{-i\delta}) = 2 - 2\cos\delta$ and $(1 - Re^{i\delta})(1 - Re^{-i\delta}) = (1 + R^2) - 2R\cos\delta$, we obtain the following expressions for the intensity of the reflected light, I_r, and transmitted light, I_t:

$$I_r = A_r A_r^* = \frac{R(2 - 2\cos\delta)}{(1 + R^2) - 2R\cos\delta} I_i = \frac{4R\sin^2(\delta/2)}{(1 - R)^2 + 4R\sin^2(\delta/2)} I_i \tag{5.42}$$

$$I_t = A_t A_t^* = \frac{T^2}{(1 + R^2) - 2R\cos\delta} I_i = \frac{(1 - R)^2}{(1 - R)^2 + 4R\sin^2(\delta/2)} I_i \tag{5.43}$$

Here, $I_i = A_0^2$ is the incident intensity, and we used the mathematical identity $\cos\delta = 1 - 2\sin^2(\delta/2)$.

From Eqs. (5.42) and (5.43), we obtain the following expression for the power reflection (reflectance) and transmission (transmittance) of the etalon:

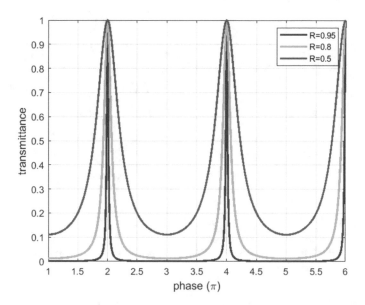

Fig. 5.5 Fabry–Perot etalon transmittance

$$\frac{I_r}{I_i} = \frac{4R\sin^2(\delta/2)}{(1-R)^2 + 4R\sin^2(\delta/2)} \tag{5.44}$$

$$\frac{I_t}{I_i} = \frac{(1-R)^2}{(1-R)^2 + 4R\sin^2(\delta/2)} \tag{5.45}$$

Note that $I_r/I_i + I_t/I_i = 1$, which indicates the conservation of the optical energy.

Figure 5.5 plots the transmittance of a Fabry–Perot etalon as a function of the round-trip phase δ for three mirror-reflectances R. The horizontal axis is the round-trip phase in the unit of π. When the round-trip phase is an integer multiple of 2π, the transmittance is at the maximum of 1 as the resonance condition is satisfied.

5.3.2 Passive Resonator

Figure 5.6 illustrates the concept of a passive resonator configured with two curved mirrors. An input laser beam enters the first mirror from the right. The mode-matching lens converts the input laser beam's profile to a resonant mode to the resonator. Under a resonant condition, the transmission of the resonator is highest, or the reflection is lowest. This situation corresponds to a peak of the transmittance plot of the Fabrey–Perot etalon, Fig. 5.5, discussed in the preceding section.

As discussed in Sect. 3.3.4, a Hermite–Gaussian wave has longitudinal and transverse modes. Under a resonant condition, the input beam must match the resonator with longitu-

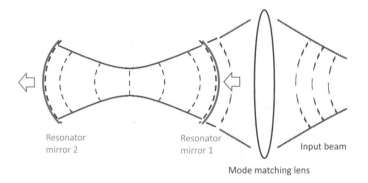

Fig. 5.6 Passive resonator

dinal and transverse modes. Below, we discuss longitudinal and transverse mode-matching for a passive resonator separately.

Longitudinal mode

According to Eq. (5.44), the condition that minimizes the reflection is $\sin(\delta/2) = 0$, or

$$\delta/2 = q\pi \tag{5.46}$$

Here, δ is the phase that the wave gains every round-trip, and q is an integer. Since the wave gains a phase of 2π every wavelength λ, we can express the round-trip phase with the resonator length l, wave velocity (speed of light) c, and frequency ν as follows:

$$\delta = 2\pi \frac{2l}{\lambda} = \frac{4\pi l}{c/\nu} \tag{5.47}$$

Using Eq. (5.47), we can express the resonant condition (5.46) in terms of the frequency as follows:

$$\nu_q = q\frac{c}{2l} \tag{5.48}$$

The resonance condition (5.48) holds as long as q is an integer. Here, q is the mode number we discussed in Fig. 5.6. So, we call the resonant frequency ν_q the qth longitudinal mode frequency. The next resonant frequency corresponding to mode number $q + 1$ is

$$\nu_{q+1} = (q + 1)\frac{c}{2l} \tag{5.49}$$

From Eqs. (5.48) and (5.49), we find the longitudinal mode's frequency interval $\Delta\nu$ as follows:

$$\Delta\nu = (q + 1)\frac{c}{2l} - q\frac{c}{2l} = \frac{c}{2l} \tag{5.50}$$

Equation (5.50) indicates that on the frequency axis, the distance between the neighboring transmission peaks of a resonator's transmittance plot like Fig. 5.5 is $\Delta \nu$. When the resonator is 10 cm long, as an example, $\Delta \nu = 3.0 \times 10^8/(2 \times 10^{-1}) = 1.5 \times 10^9 = 1.5$ GHz. If the input wave's wavelength is 500 nm (green laser), its frequency is $3.0 \times 10^8/500 \times 10^{-9} = 6.0 \times 10^{14}$ Hz. We find that in an optical resonator, the longitudinal modes are packed. This observation is not surprising if we note that the resonator length of 10 cm is 20,000 times greater than the wavelength of 500 nm.

Finesse

The Finesse [14] of a resonator is a measure of the energy loss. We can evaluate the energy loss using the time constant t_c defined as follows:

$$\frac{d\epsilon}{dt} = -\frac{\epsilon}{t_c} \tag{5.51}$$

Here, ϵ is the energy stored in the resonator. Solving the differential equation (5.51), we find that the stored energy decays exponentially.

$$\epsilon(t) = \epsilon_0 e^{-t/t_c} \tag{5.52}$$

Here, ϵ_0 is the initial intra-resonator energy at $t = 0$. In the time domain, the intra-resonator energy decays in the same fashion as the damped harmonic oscillation discussed in Fig. 1.2. We call this oscillatory-decaying pattern a ring-down. If we monitor the output of a passive resonator after cutting the input beam, we can observe a ring-down pattern on which we can evaluate t_c as the time when the energy decays by a factor of e.

If the energy loss per trip from one mirror to the other is L, we can express it as follows:

$$\frac{d\epsilon}{dt}\left(\frac{l}{c}\right) = -L\epsilon \tag{5.53}$$

Here, l/c is the trip time, and the left-hand side of Eq. (5.53) is the energy decay rate times the trip time, i.e., the energy reduction per trip. The right-hand side indicates that the energy reduction per trip is the loss rate L times the energy ϵ.

From Eqs. (5.51) and (5.53), we can relate the time constant to the loss rate L.

$$t_c = \frac{l}{c}\frac{1}{L} \tag{5.54}$$

Equation (5.54) tells us the time constant is proportional to the trip time l/c and inversely proportional to the loss rate.

If the mirror transmittance T dominates the loss L, we can replace L in the denominator of Eq. (5.54) with $1 - R$. Here, R is the mirror reflectance. In this case, the time constant becomes as follows:

$$t_c = \frac{l}{c(1-R)} \tag{5.55}$$

Above, we considered energy loss in a resonator in the time domain. In the frequency domain, we can discuss the energy loss in conjunction with the quality factor and spectral width, and relate these parameters to Finesse. The quality factor (Q-factor) of an oscillator is defined as follows:

$$Q = \frac{\omega \epsilon}{P} = \frac{\omega \epsilon}{d\epsilon/dt} \tag{5.56}$$

Here, ω is the angular frequency and P the power. Using Eq. (5.52), we can express Q as follows:

$$Q = \omega t_c \tag{5.57}$$

We find the FWHM (Full Width at Half Maximum) of a spectrum $\Delta v_{1/2}$ is related to Q.

$$\Delta v_{1/2} = \frac{v}{Q} = \frac{v}{\omega t_c} = \frac{1}{2\pi t_c} \sim \frac{c(1-R)}{2\pi l \sqrt{R}} \tag{5.58}$$

Here, we use Eq. (5.55) in the last equal sign.

Finally, the Finesse is defined as follows:

$$F \equiv \frac{\pi \sqrt{R}}{1-R} \tag{5.59}$$

Using Eq. (5.58), we can express the Finesse with Q or t_c.

$$F = \frac{c}{2l\Delta v_{1/2}} = \frac{\lambda Q}{2l} = \frac{\pi c t_c}{l} \tag{5.60}$$

Equation (5.60) indicates that the Finesse is proportional to the Q-factor or time constant.

Transverse mode

Consider the phase term of the Hermite–Gaussian solution we derived in Chap. 3.

$$E_{m,n} = E_0 \frac{w_0}{w(z)} H_m \left(\frac{\sqrt{2}x}{w(z)} \right) H_n \left(\frac{\sqrt{2}y}{w(z)} \right) exp \left[-\frac{x^2+y^2}{w^2(z)} \right]$$
$$exp \left[-\frac{ik(x^2+y^2)}{2R(z)} - ikz + i(m+n+1)\phi_G \right] \tag{5.61}$$

The phase term ikz measures the change in phase as the wave propagates along the propagation axis. This on-axis phase change is the same as a plane wave of the same wavelength. The phase factor ϕ_G, known as the Gouy phase shift, represents the on-axis phase shift from that for a plane wave.

$$\phi_G = \tan^{-1} \frac{\lambda z}{\pi w_0^2} \tag{5.62}$$

Here, λ is the wavelength, and w_0 is the beam waist size of the Gaussian beam. The indices m and n in the term $(m + n + 1)$ in front of the Gouy phase shift represent the order of the Hermite function, H_m and H_n. The $i(m + n + 1)\phi_G$ term indicates the on-axis phase changes for the transverse modes of the mth and nth orders.

Notice that m or n can take any value independent of the on-axis phase ikz, indicating that a light wave can have multiple transverse modes (m, n) for the same longitudinal modes q. In other words, a resonator of a fixed length l corresponding to a single longitudinal mode can take multiple transverse modes. Below, we discuss this situation using a resonator of a two-mirror configuration as an example.

Consider that the two mirrors are at $z = z_1$ and $z = z_2$. The resonant condition yields the following equation:

$$kl = q\pi \tag{5.63}$$

Here, l is the resonator length, $k = 2\pi/\lambda$, and q is the longitudinal mode number. Calling the phase at $z = z_1$ and $z = z_2$ $\theta(z_1)$ and $\theta(z_2)$, we can write the resonant condition (5.63) as follows:

$$\theta(z_2) - \theta(z_1) = q\pi \tag{5.64}$$

Using the phase expression used in the last term of Eq. (3.102), we can derive the following equation from Eq. (5.64):

$$kl - (m + n + 1)\left[\tan^{-1}\frac{z_2}{z_R} - \tan^{-1}\frac{z_1}{z_R}\right] = q\pi \tag{5.65}$$

Here, z_R is the Rayleigh length.

$$z_R = \frac{\pi w_0^2}{\lambda} \tag{5.66}$$

Since the neighboring longitudinal modes satisfy the following equation (Eq. 5.63),

$$k_{q+1} - k_q = \frac{\pi}{l} \tag{5.67}$$

Thus, we obtain the following set of equations:

$$k_1 l - (m + n + 1)_1\left[\tan^{-1}\frac{z_2}{z_R} - \tan^{-1}\frac{z_1}{z_R}\right] = q\pi \tag{5.68}$$

$$k_2 l - (m + n + 1)_2\left[\tan^{-1}\frac{z_2}{z_R} - \tan^{-1}\frac{z_1}{z_R}\right] = q\pi \tag{5.69}$$

By subtracting Eq. (5.69) from Eq. (5.68), we find the following:

$$(k_1 - k_2)l = [(m + n + 1)_1 - (m + n + 1)_2]\left(\tan^{-1}\frac{z_2}{z_R} - \tan^{-1}\frac{z_1}{z_R}\right) \tag{5.70}$$

Defining the change in the transverse mode as $\Delta(m + n)$, we can express the corresponding transverse mode frequency interval as follows:

Fig. 5.7 Confocal resonator
and longitudinal and transverse
mode frequency separations

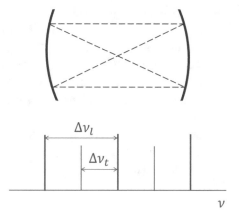

$$\Delta v_t = \frac{c}{2\pi l}\Delta(m+n)\left(\tan^{-1}\frac{z_2}{z_R} - \tan^{-1}\frac{z_1}{z_R}\right) \tag{5.71}$$

As an example, consider a confocal resonator in Fig. 5.7. The positions of the two mirrors of the same curvature are $z_1 = -z_R$ and $z_2 = z_R$. In this case, $\tan^{-1}(z_2/z_R) = -\tan^{-1}(z_1/z_R) = \pi/4$. Using this condition in Eq. (5.71), we obtain the following expression for the transverse mode frequency interval:

$$\Delta v_t = \frac{c}{2\pi l}\Delta(m+n)\left(\frac{\pi}{4} - (-\frac{\pi}{4})\right) = \frac{1}{2}\frac{c}{2l}\Delta(m+n) \tag{5.72}$$

Equation (5.50) indicates the longitudinal mode frequency separation is $\Delta v_l = c/(2l)$. So, from Eq. (5.72) in the case of a confocal resonator, we find that the transverse mode frequency is at a longitudinal mode frequency or at the middle of a neighboring pair of longitudinal mode frequencies. The insert in Fig. 5.7 illustrates the situation.

5.3.3   Active Resonator

Threshold condition

By placing a gain medium inside an optical resonator we can configure an active resonator, which we commonly call a laser. Refer to Fig. 5.8. Let l be the resonator length, r_1 and r_2 be the reflectivity of the left and right mirrors, γ and α be the gain and loss per single pass. Consider that a light wave reflected by the left mirror travels to the right mirror and gets reflected there again. Remembering the exponential gain or loss discussed with Eq. (5.26), we can express the net gain as $r_1 r_2 exp(\gamma - \alpha)l)$. By setting the net gain equal to unity, we can obtain the following equation to express the situation where the gain balances the loss. We call this condition the threshold condition. Under this condition, the optical field

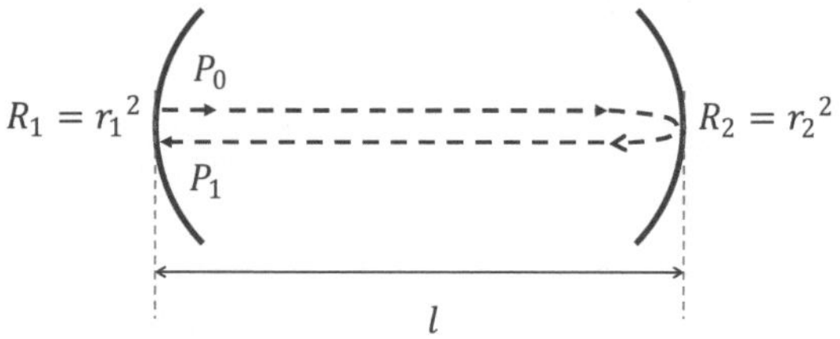

Fig. 5.8 Intra-resonator power
before and after one round trip

establishes a steady state.

$$r_1 r_2 e^{(\gamma - \alpha)l} = 1 \tag{5.73}$$

From Eq. (5.73), we obtain the following expression for the gain for the steady state:

$$\gamma_t = \alpha - \frac{1}{l} \ln(r_1 r_2) \tag{5.74}$$

Here, the subscript t denotes "threshold". We call γ_t the the threshold gain.

From Eqs. (5.25) and (5.74), we obtain the following expression for the threshold population inversion:

$$N_t = (N_2 - N_1)_t = \frac{8\pi \nu^2 t_{spont}}{c^2 g(\nu)} \gamma_t = \frac{8\pi \nu^2 t_{spont}}{c^2 g(\nu)} \left(\alpha - \frac{1}{l} \ln(r_1 r_2) \right) \tag{5.75}$$

Consider the change in optical energy after one round trip in a resonator in Fig. 5.8. Call the optical power of the light just leaving the left mirror P_0 and that of the light leaving the same mirror after one round trip P_1. The reflectance (power reflectivity) of the two mirrors are, respectively, R_1 and R_2. We can relate P_1 and P_0 as follows:

$$P_1 = P_0 e^{-\alpha l} R_2 e^{-\alpha l} R_1 = P_0 R_1 R_2 e^{-2\alpha l} \tag{5.76}$$

From Eq. (5.76), we find the power loss per round trip as follows:

$$P_0 - P_1 = P_0 \left(1 - R_1 R_2 e^{-2\alpha l} \right) \tag{5.77}$$

In other words, the optical power loses the fraction $\left(1 - R_1 R_2 e^{-2\alpha l} \right)$ in every round trip. Since it takes $2l/c$ s for each round trip, we can express the rate of energy loss W_{loss} as follows:

$$W_{loss} = \frac{1}{t_c} = \frac{1 - R_1 R_2 e^{-2\alpha l}}{\frac{2l}{c}} = -\frac{c}{2l} \left(R_1 R_2 e^{-2\alpha l} - 1 \right) \tag{5.78}$$

Assume $R_1 R_2 e^{-2\alpha l} \cong 1$, i.e., the loss per round trip is small or $|x - 1| << 1$. Taylor expansion of function $f(x)$ around $x = 1$ is as follows:

$$f(x) = f(1) + f'(1)(x - 1) + \frac{1}{2} f''(x - 1)^2 + \cdots \tag{5.79}$$

Fig. 5.9 Normalized $g(\nu)$ and linewidth

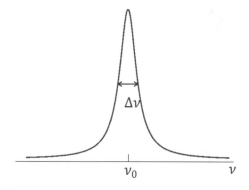

Applying Eq. (5.79) to the natural logarithmic function and taking the linear term only, we find the following approximation:

$$\ln(x) \cong \ln(1) + (x - 1) = x - 1 \tag{5.80}$$

Using approximation (5.80), we can rewrite Eq. (5.78) as follows:

$$\frac{1}{t_c} = -\frac{c}{2l}\left(R_1 R_2 e^{-2\alpha l} - 1\right) = -\frac{c}{2l}\ln\left(R_1 R_2 e^{-2\alpha l}\right)$$
$$= -\frac{c}{2l}\left(\ln(R_1 R_2) + \ln e^{-2\alpha l}\right) = -\frac{c}{2l}\left[\ln(R_1 R_2) - 2\alpha l\right] \tag{5.81}$$

Rewriting the power reflectance using amplitude reflection coefficients as $R_1 R_2 = (r_1 r_2)^2$, we can put Eq. (5.81) in the following form:

$$\frac{1}{t_c} = -\frac{c}{2l}\left[2\ln(r_1 r_2) - 2\alpha l\right] = c\left(\alpha - \frac{1}{l}\ln(r_1 r_2)\right) \tag{5.82}$$

From Eqs. (5.75) and (5.82), we find the following expression for the threshold population inversion:

$$N_t = \frac{8\pi \nu^2 t_{spont}}{c^2 g(\nu)}\left(\alpha - \frac{1}{l}\ln(r_1 r_2)\right) = \frac{8\pi \nu^2 t_{spont}}{c^3 t_c}\frac{1}{g(\nu)} \tag{5.83}$$

In Eq. (5.83), $g(\nu)$ is a normalized Lorentz distribution (Eq. (5.20)). Figure 5.9 illustrates the spectral $g(\nu)$ and its FWHM.

We can approximate the area of $g(\nu)$ in Fig. 5.9 as a triangle of the base 2Δ and area 1. So, we can approximate the height $g(\nu_0) = g_{peak}$ as follows:

$$S = \frac{1}{2}(2\Delta\nu)g_{peak} = \Delta\nu g_{peak} = 1$$

Hence,

$$g(\nu_0) \cong \frac{1}{\Delta\nu} \tag{5.84}$$

With expression (5.84), we can write the threshold population inversion (5.83) as follows:

$$N_t = \frac{8\pi v^2 t_{spont}}{c^3 t_c} \frac{1}{g(v)} = \frac{8\pi v^2 t_{spont}}{c^3 t_c} \Delta v \tag{5.85}$$

Expression (5.85) is convenient to find an approximate threshold population inversion from the linewidth Δv, resonator decay time t_c, and radiative lifetime t_{spont}.

Furthermore, according to Eq. (5.5) $8\pi h v^3 / c^3$ is the optical energy density per frequency, ρ_v (see Eqs. (5.5) and (5.14)). So, we can write Eq. (5.85) in the following form as well:

$$N_t(hv)\omega_{21} = \frac{N_t(hv)}{t_{spont}} = \frac{8\pi h v^3}{c^3 t_c} \Delta v = \frac{\rho_v \Delta v}{t_c} \tag{5.86}$$

Here, in going through the first equal sing of Eq. (5.86), we replace the decay rate ω_{21} with the reciprocal of the spontaneous emission lifetime t_{spont}. Thus, we can view the quantity $N_t(hv)\omega_{21}$ as the optical energy within the linewidth under the threshold population inversion divided by the resonator decay time t_c, i.e., the spontaneous emission power density, p_s, under the threshold population inversion.

$$p_s = N_t(hv)\omega_{21} \tag{5.87}$$

Gain spectral shape and longitudinal mode line spectrum

Remember in Sect. 5.2.3, we discussed that atomic transitions are Lorentzian (Eq. 5.21). Equation (5.45) derived in Sect. 5.3.1 shows the transmission characteristic of an optical resonator is also Lorentzian. (The numerators of these expressions have different forms, but since the numerator determines the peak value of the function, both expressions represent the Lorentz distribution.) These facts indicate that an active resonator is a combination of two Lorentzian, the first represents the emission spectrum of the gain medium, and the second the resonator characteristics. Figure 5.10 illustrates the situation schematically. Here, the dashed line is the emission spectrum, and the solid lines are spectral shapes of three neighboring longitudinal modes. So long as a longitudinal mode's spectrum is within the central region of the gain spectrum, laser oscillation can occur.

As illustrated by Fig. 5.45, the resonator line shape is normally narrower than the gain spectrum. Consider this situation using a He–Ne (Helium–Neon) laser [15] resonator oscillating at the wavelength of 632.8 nm with a resonator length of 20 cm as an example. From Eq. (5.48), we find the fundamental (lowest) resonant frequency is 7.5×10^8 Hz $= 0.75$ GHz.

$$v_1 = v_q|_{q=1} = q\frac{c}{2l} = \frac{3 \times 10^8}{2 \times 0.2} = 7.5 \times 10^8 \tag{5.88}$$

Fig. 5.10 Atomic transition spectral width resonator line width

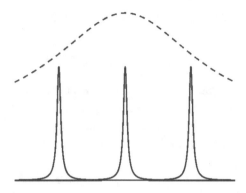

The central frequency of the 632.8 nm emission spectrum is 4.7408×10^{14} Hz $= 474.08$ THz.

$$\nu_{\text{HeNe}} = \frac{3 \times 10^8}{632.8 \times 10^{-9}} = 4.7408 \times 10^{14} \qquad (5.89)$$

Dividing ν_{HeNe} by ν_1, we find that resonator's 632111th harmonics would resonate with the central frequency of the He–Ne laser's gain spectrum. In other words, the mode number $q = 632111$ is the closest to the central frequency of the He–Ne laser's gain spectrum.

The linewidth of the He–Ne's emission spectrum centered at 632.8 nm (the gain width) is approximately 1 GHz, $\Delta \nu_{\text{HeNe}} \cong 1 \times 10^9$ Hz. The longitudinal mode's frequency separation ($\Delta \nu_l c / (2l)$, see Eq. (5.50)) is 0.75 GHz. Since $\Delta \nu_l < \Delta \nu_{\text{HeNe}}$ two longitudinal mode frequencies are within the gain width, indicating that mode numbers $q = 632110$ or $q = 632112$ can oscillate. When the threshold population inversion reaches the threshold value at one of these modes, the laser oscillates at the corresponding frequency. (In general, if the lasing species is inhomogeneously broadened, the laser can oscillate at multiple longitudinal modes within the gain width. See Chap. 6 of [10] for homogeneous and inhomogeneous broadenings.)

Laser operation

Recall that we derived the expression of the threshold population inversion $N_t \equiv (N_2 - N_1)_t$, Eq. (5.85), from the expression of the threshold gain γ_t, Eq. (5.74), which states that the threshold gain is equal to the intra-resonator loss per unit length and transmission loss through the mirrors ($R_1 R_2 \neq 1$). Since the threshold population inversion is proportional to the threshold gain (Eq. (5.85)), an increase in the population inversion corresponds to the situation where the gain is greater than the loss. Under that condition, the intra-cavity optical power increases after every round trip, which breaks the steady-state condition.

This observation indicates that under a steady-state condition with a given set of α, R_1, and R_2, the population inversion is locked at the threshold value. Thus, from Eq. (5.34), we find that the following equality always holds:

$$N_t = \frac{R}{W_i + \omega_{21}} \tag{5.90}$$

We can rationalize that the threshold population inversion is locked at the threshold value by the following argument. In Eq. (5.90), if the pumping rate to the upper state (R_2 in Eq. (5.34)) increases, the stimulated emission rate W_i increases in the denominator keeping the value of N_t the same, i.e., the threshold condition is kept as the stimulated emission rate W_i increases with the pumping rate R.

Using Eq. (5.90), we can discuss the relationship between the pumping rate and the optical power stored in the resonator. Solve Eq. (5.90) for the induced emission rate W_i.

$$W_i = \frac{R}{N_t} - \omega_{21} \tag{5.91}$$

The optical power density due to the induced emission, p_e, is the product of the emission rate, population inversion (number density), and photon energy. Thus,

$$p_e = N_t W_i (h\nu) \tag{5.92}$$

Substituting Eq. (5.91), we can rewrite Eq. (5.92) as follows:

$$p_e = N_t \omega_{21} (h\nu) \left(\frac{R}{N_t \omega_{21}} - 1 \right) \tag{5.93}$$

Remembering the spontaneous power expression under the threshold population inversion (5.87), we can write Eq. (5.93) as follows:

$$p_e = p_s \left(\frac{R}{p_s / h\nu} - 1 \right) = p_s \left(\frac{R}{R_t} - 1 \right) \tag{5.94}$$

Equation (5.94) indicates that the induced optical power density increases as the pumping rate R increases from the threshold pumping rate R_t. In other words, if the pumping rate is at the threshold level, the induced emission is just to compensate for the intra-resonator loss and we cannot take the optical power out from the resonator to use as laser output power.

Optimum Power Out

Since the population inversion is locked at the threshold value regardless of the pumping strength, from (5.25), we obtain the following expression for the gain coefficient γ:

$$\gamma = N_t \frac{c^2 g(\nu)}{8\pi \nu^2 t_{spont}} = \frac{R}{W_i + \omega_{21}} \frac{c^2 g(\nu)}{8\pi \nu^2 t_{spont}} \tag{5.95}$$

Here, we use Eq. (5.34) for N_t. When the intra-resonator emission is low, $W_i << \omega_{21}$. Under this condition, the gain per length becomes as follows:

$$\gamma_0 = \frac{R}{\omega_{21}} \frac{c^2 g(v)}{8\pi v^2 t_{spont}} \tag{5.96}$$

We call the total gain for the resonator length l corresponding to γ_0 the small-signal gain g_0.

$$g_0 = \gamma_0 l \tag{5.97}$$

Equations (5.95) and (5.96) yield the following equation:

$$\frac{\gamma}{\gamma_0} = \frac{\omega_{21}}{W_i + \omega_{21}} = \frac{1}{W_i/\omega_{21} + 1} \tag{5.98}$$

From Eqs. (5.87) and (5.92), we find as follows:

$$\frac{W_i}{\omega_{21}} = \frac{p_e}{p_s} \tag{5.99}$$

Using Eqs. (5.98) and (5.99), we obtain the following expression for p_e:

$$p_e = p_s \left(\frac{\gamma_0}{\gamma} - 1\right) \tag{5.100}$$

Here, γl is the total gain with $W_i \neq 0$, and from Eq. (5.97), $\gamma_0 = g_0/l$. Since γl balances the total loss L, $\gamma = L/l$. We can express p_e using gain per loss as follows:

$$p_e = p_s \left(\frac{g_0/l}{L/l} - 1\right) = p_s \left(\frac{g_0}{L} - 1\right) \tag{5.101}$$

The total loss L consists of residual loss (absorption by impurities etc) and the mirror's transmittance T.

$$L = L_i + T \tag{5.102}$$

Substitution of Eq. (5.102) into Eq. (5.101) yields the following expression of p_e:

$$p_e = p_s \left(\frac{g_0}{L_i + T} - 1\right) \tag{5.103}$$

Here, p_e is the optical power inside the resonator balancing the total loss L. Therefore, $T/(L_i + T)$ portion of p_e corresponds to the optical power density that the resonator loses via the mirror's transmission. In other words, p_{out} expressed by the following expression is the optical power density coming out of the resonator through the mirror, i.e., the output power density.

$$p_{out} = p_e \frac{T}{L_i + T} = p_s \left(\frac{g_0}{L_i + T} - 1\right) \frac{T}{L_i + T} \tag{5.104}$$

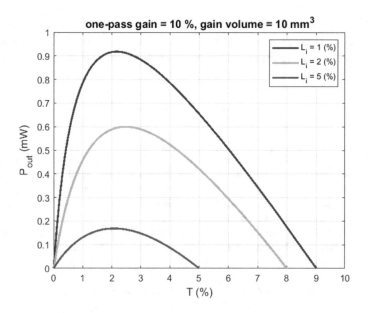

Fig. 5.11 T vs output coupling

Equation (5.104) indicates that depending on the residual loss L_i, there is an optimum mirror transmittance that maximizes the output power density.

Since $1/t_c$ is the loss rate (See Eq. (5.78)),

$$\frac{1}{t_c} = \frac{L_i + T}{l/c} \tag{5.105}$$

Here, l/c is the time for one pass, and the total loss L divided by this time is the loss rate. Using Eq. (5.85) for N_t in Eq. (5.87), we can express p_s as follows:

$$p_s = \frac{8\pi h\nu^3 t_{spont}}{c^3 t_c}\Delta\nu\omega_{21} = \frac{8\pi h\nu^3}{c^2 l}\Delta\nu(L_i + T) \tag{5.106}$$

Substituting Eq. (5.106) into p_s, we can rewrite Eq. (5.104) in the following form:

$$p_{out} = \left(\frac{8\pi h\nu^3 \Delta\nu}{c^2 l}\right)\left(\frac{g_0}{L_i + T} - 1\right)T \tag{5.107}$$

By finding T that makes $\partial p_{out}/\partial T = 0$, we find that the following transmittance maximizes p_{out}:

$$T_{opt} = -L_i + \sqrt{g_0 L_i} \tag{5.108}$$

Note that Eq. (5.107) evaluates the power density in W/m^3. To find the output intensity, we need to multiply the gain volume. Figure 5.11 plots sample calculations to show the relation between T and p_{out} using the above He–Ne laser case with a single-pass gain of 10 % and a gain volume of 10 mm^3.

References

1. M. Bertolotti, *The History of the Laser*, (CRC Press, Boca Raton, New York, 2018) pp. 101–114
2. https://www.rp-photonics.com/amplified_spontaneous_emission.html
3. D. Griffiths, *Introduction to Quantum Mechanics*, (Pearson Education, Ltd. London, 2003)
4. L. I. Schiff, *Quantum Mechanics* International Student edn. (McGrow-Hill, New York, 1968)
5. D. Lindley, *Uncertainty* (Anchor Books, New York, 2008)
6. 1.9: The Heisenberg Uncertainty Principle, LibreTexts, https://chem.libretexts.org/Courses/ BethuneCookman_University/B-CU%3ACH-331_Physical_Chemistry_I/CH-331_Text/ CH-331_Text/01%3A_The_Dawn_of_the_Quantum_Theory/1.9%3A_The_Heisenberg_ Uncertainty_Principle
7. D. Griffiths, *Introduction to Quantum Mechanics*, (Pearson Education, Ltd. London, 2003), pp. 30–40
8. P. Ewart, Spectroscopy| Nonlinear Laser Spectroscopy, in *Encyclopedia of Modern Optics* (Elsevier Ltd., Amsterdam, 2005) pp. 109–119
9. Planck's Radiation law, http://gdckulgam.edu.in/Files/f07ef270-7e91-4716-8825- 2966f17cc0f7/Menu/Plancks_Radiation_law_3da32a73-3848-4135-bd19-e110bd2dfdbd. pdf (accessed on Aug 11, 2022)
10. A. Yariv, *Introduction to Optical Electronics* (Holt, Rinehart and Winston, New York, USA, 1971)
11. G. G. Lister, J. F. Waymouth, III.A.2 Local Thermal Equilibrium, in *Encyclopedia of Physical Science and Technology* 3rd edn. (Elsevier Ltd., Amsterdam, 2003) pp. 557–595
12. C. K. Rhodes, Ed., *Excimer Lasers*, (Springer, Berlin, Heidelberg, 1979)
13. R. Pachotta, Fabry–Pérot Interferometers, RP Photonics Encyclopedia, https://www.rp-photonics.com/beam_divergence.html (accessed on August 11, 2022)
14. R. Pachotta, Finesse, RP Photonics Encyclopedia, https://www.rp-photonics.com/beam_ divergence.html (accessed on August 11, 2022)
15. L. W. Anderson, J. B. Boffard, *Lasers for Scientists and Engineers* (World Scientific, Singapore, 2017), pp. 233-253 Chap. 10: The Helium-Neon Laser https://doi.org/10.1142/9789813224308_ 0010

Appendix A
Maxwell's Term

Maxwell added the first term as he noted that without this term some inconsistency would occur in the situation depicted in Fig. A.1 where the ends of a pair of wires are configured as forming a parallel plate capacitor. Consider a closed-loop C around the wire, and two surfaces S_1 and S_2. Here S_1 cut the wire, and S_2 encloses the capacitor plate without cutting the wire. According to Stoke's theorem, the following equation holds:

$$\int_C \mathbf{B} \cdot d\mathbf{l} = \int \int_S \nabla \times \mathbf{B} \cdot d\mathbf{S} \tag{A.1}$$

Here the left-hand side of Eq. (A.1) is the line integral of the magnetic field $\mathbf{B}$ along loop C, and the right-hand side is the surface integral of $\nabla \times \mathbf{B}$ over the surface area S. If we apply Ampère's law, $\nabla \times \mathbf{B}$ is equal to the density of conduction current $\mu\mathbf{j}$. So, the right-hand side of Eq. (A.1) is the total conduction current penetrating through surface μJ.

$$\int \int_S \nabla \times \mathbf{B} \cdot d\mathbf{S} = \mu J \tag{A.2}$$

Apparently, the value of the left-hand side of Eq. (A.1) depends only on loop C, independent of the surface. However, the value of the right-hand side of Eq. (A.1) depends on the choice of the surface. If we use surface S_1, the surface integral represents the total conduction current flowing through the wire and yields a finite value. So, Eq. (A.1) can hold. However, if we use surface S_2 there is no conduction current flowing through the surface and the right-hand side of Eq. (A.1) becomes null. Since the left-hand side of Eq. (A.1) takes a finite value, Eq. (A.1) cannot hold.

Maxwell's term solves this problem as follows. Take divergence of Eq. (2.2) and use the mathematical identity $\nabla \cdot (\nabla \times) = 0$. We obtain the following equation:

$$\nabla \cdot \epsilon \frac{\partial \mathbf{E}}{\partial t} = -\nabla \cdot \mathbf{j} \tag{A.3}$$

© The Editor(s) (if applicable) and The Author(s), under exclusive license to
Springer Nature Switzerland AG 2023
S. Yoshida. *Fundamentals of Optical Waves and Lasers*. Synthesis Lectures on Wave
Phenomena in the Physical Sciences, https://doi.org/10.1007/978-3-031-18188-7

(a)   (b)

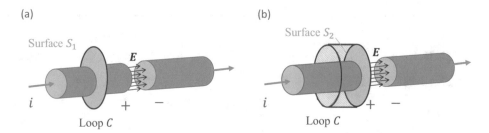

Fig. A.1 Ends of a pair of conductors form a parallel plate capacitor. Closed-loop C is used on the left-hand side of Eq. (A.1). In **a** the surface used for the right-hand side of Eq. (A.1) is the flat circle surrounded by C. In **b** the surface for the right-hand side of Eq. (A.1) is in a bucket-like shape. The bottom of the bucket crosses the electric field but not the conduction current. The side of the bucket-shaped surface is parallel to the conduction current and the electric field and therefore irrelevant to the magnetic field generation

Substitution of Eq. (2.1) into the left-hand side of Eq. (A.3) leads to the following form of the charge conservation, which explicitly indicates that the electric charges are not generated or destroyed.

$$\frac{\partial \rho}{\partial t} = -\nabla \cdot \mathbf{j} \tag{A.4}$$

Equation (A.4) indicates that when the charge density varies with time in a certain volume in space, there is necessarily current flows accompanied. When the charge density increases, the net current flows into the volume, and when the charge density decreases the net current flows out of the space. This represents that the current is a flow of charges and the charges are never generated or destroyed.

By defining the displacement current J_d as follows we can view the charge conservation law as representing the continuity of the current at the capacitor plate.

$$J_d = \epsilon \int \int_S \frac{\partial \mathbf{E}}{\partial t} \cdot d\mathbf{S} \tag{A.5}$$

By interpreting Maxwell's term as the density of the displacement current, j_d, we can rewrite Eq. (A.3) in the following form:

$$\nabla \cdot \mathbf{j} = -\nabla \cdot \mathbf{j_d} \tag{A.6}$$

Equation (A.6) indicates that the flow-out of conduction current at the capacitor plate (the left-hand side of Eq. (A.6)) is equal to the flow-in of the displacement current (the right-hand side) into the capacitor gap. In other words, inside the wire, the conduction current carries the electric charges. When the current comes to the capacitor plate, it cannot flow in the same fashion as before. Inside the capacitor, the displacement current flows so that the charge conservation is guaranteed.

Appendix B
Management of Optical Polarization

In optical experiments often the laser's polarization becomes important. For instance, laser beams having mutually orthogonal linear polarization do not interfere with each other because the electric fields are orthogonal. Therefore, they cannot be used for an interferometer's interfering lights. In other situations, we often want to prevent reflected lights to disturb the sensitive optical configuration. In an extreme case, reflection of a high-power laser back into the laser resonator degrades the performance of the laser resonator or even can damage the gain medium. In this situation, we need to isolate the laser source from the rest of the optical setup. Below, we discuss some techniques to control the polarization of laser beams.

B.1   Half-Wave Plate

A half-wave plate [1] is commonly used to rotate a linear polarization for a given angle. Figure B.1 illustrates how a half-wave plate turns the orientation of a linearly polarized light. The material of this half-wave plate is a birefringent crystal. Here, n_e and n_o denote, respectively, the index of refraction for the extraordinary and ordinary lights. The extraordinary ray propagates more slowly (due to the greater index of refraction) than the ordinary ray. So, the n_o corresponds to the fast axis discussed.

As we discussed in Fig. 3.7, when a linearly polarized light passes through a birefringent crystal the polarization is altered. Half-wave plates utilize this phenomenon by adjusting the crystal thickness so that after passing through the plate, the phase along the fast axis advances by half-period (2π in phase). Therefore, at the exit end of the half-wave plate the orientation of the electric field vector is rotated symmetrically about the slow axis (n_e) as the inserts of Fig. B.1. Here ϕ denotes the angle of the incident polarization to the slow axis. Notice that the total angle of rotation is 2ϕ and we can choose any angle for ϕ. If we want to rotate the incident polarization by α, all we need to do is set $\phi = \alpha/2$.

S. Yoshida, *Fundamentals of Optical Waves and Lasers*, Synthesis Lectures on Wave
Phenomena in the Physical Sciences, https://doi.org/10.1007/978-3-031-18188-7

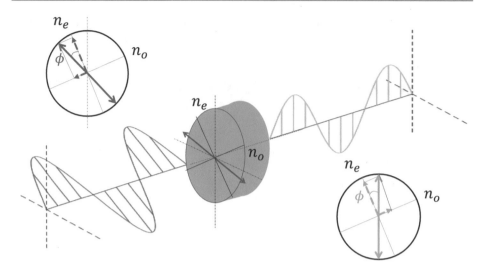

Fig. B.1 Rotation of linear polarization with half-wave plate

B.2   Beam Attenuation

One of the simplest and most useful applications of polarization rotation with a half-wave plate is beam attenuation. By placing a polarizer after a half-wave plate (see Fig. B.2a), we can block part of the optical power. For instance, assume that the polarizer passes through vertical polarization and the laser source outputs a vertically polarized beam with amplitude E_0. By rotating the original polarization by angle α to the vertical axis, only $E_0 \cos \alpha$ passes through the polarizer. Figure B.2a illustrates such a setup.

Alternatively, we can use a simple glass plate in place of a polarizer. Depending on the polarization of the incident beam and angle of incidence, we can control the portion of reflected and transmitted optical power using the external reflection plot Fig. 4.5.

This is an easy configuration but you need to be careful with the unused portion of the optical power. If you use a Glan–Foucault polarizer, for instance, and orient the half-wave plate so that the polarization transmitted through the polarizer to be used, the unused polarization is kicked to a side as shown in Fig. 4.7. If you do not block this unused portion of the optical power, it can go anywhere as a stray beam and can hurt you or other instruments. When you are using an infrared high-power laser extra caution is needed as the stray beam is invisible and can be high in power.

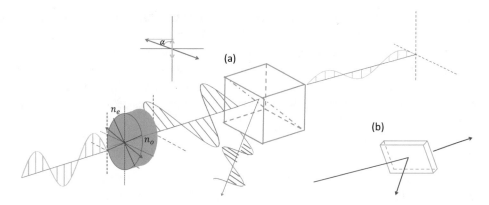

Fig. B.2 Beam attenuation with a half-wave plate and **a** polarizer or **b** glass plate

---

B.3   Faraday Isolator

A Faraday rotator rotates a linearly polarized light by 45°. Figure B.3 shows a how Faraday rotator rotates the incident polarization. In this example, the Faraday rotator rotates the incident polarization clockwise. Unlike a half-wave plate, the direction of rotation is fixed to the laboratory coordinate. In the case of this example, the Faraday rotator rotates linearly polarized light counterclockwise if it is incident from the rear side. Therefore, any light reflected back by an optical component placed downstream of the Faraday rotator becomes orthogonal to the polarization incident to the Faraday rotator (because the total rotation is $2 \times 45° = 90°$.) This situation is contrastive to a half-wave plate because as mentioned

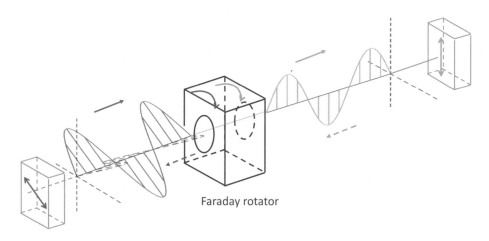

Faraday rotator

Fig. B.3 Faraday isolator

above half-wave plates rotate the polarization relative to the slow axis. If we replace the Faraday rotator with a half-wave plate in Fig. B.3, the polarization passing through the wave plate from the downstream side takes the same polarization as the incident beam.

A Faraday isolator consists of a Faraday rotator and polarizers. In the example shown in Fig. B.3, the returning beam from the downstream side is blocked by a polarizer placed before the Faraday isolator passing through the incident beam. A Faraday isolator is widely used to isolate one part of the optical system from the rest.

Reference

1. R. Pachotta, Waveplates, RP Photonics Encyclopedia, https://www.rp-photonics.com/waveplates.html (accessed on August 7, 2022)

Appendix C
Interferometry

C

C.1   Temperature/Concentration Measurement

The optical path length is defined as follows:

$$OPL = 2\pi \frac{l}{\lambda_0/n} = 2\pi \frac{nl}{\lambda_0} = nlk_0 \tag{C.1}$$

Here l is the physical optical path length, λ_0 is the wavelength in vacuum, n is the index of refraction, and $k_0 = 2\pi/\lambda$ is the wave number in a vacuum.

The index of refraction generally depends on temperature. The index of refraction of a solution depends on the concentration. Thus, from Eq. (C.1), we find that provided that the physical path length is constant the following expression represents a change in the optical path length in the unit of rad.

$$d(OPL) = dn(lk_0) = \left(\frac{\partial n}{\partial T} dT + \frac{\partial n}{\partial N} dN \right)(lk_0) \tag{C.2}$$

Here, T is the temperature and N is the concentration. By dividing Eq. (C.2) by 2π, we can express the change in the optical path length in the unit of "1/wave" as follows:

$$d(OPL)_{wave} = \left(\frac{\partial n}{\partial T} dT + \frac{\partial n}{\partial N} dN \right) \frac{l}{\lambda_0} \tag{C.3}$$

We can set up an optical interferometer and use Eq. (C.2) to measure the concentration of a solution. In principle, by precisely controlling the laboratory temperature we can find the solution concentration by measuring the change in the phase from a reference solution (such as distilled water). Often, however, the temperature fluctuation of laboratory air compromises the accuracy of the measurement. The temperature coefficient of refractive index in air is of

© The Editor(s) (if applicable) and The Author(s), under exclusive license to
Springer Nature Switzerland AG 2023
S. Yoshida. *Fundamentals of Optical Waves and Lasers*. Synthesis Lectures on Wave
Phenomena in the Physical Sciences, https://doi.org/10.1007/978-3-031-18188-7

Fig. C.1 Mach–Zehnder
interferometer for solution
concentration measurement

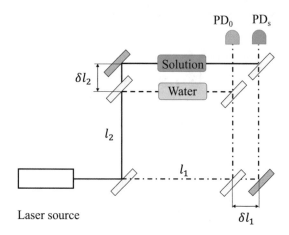

the order of 10^{-6} (1/K) [1]. This means that if the physical path length in air is 1 m, from Eq. (C.3), green light ($\lambda = 500$ nm) experiences $dnl = 10^{-6}/(500 \times 10^{-9}) = 2$ wave number per temperature fluctuation of 1 K.

An optical arrangement shown in Fig. C.1 reduces the effect of air temperature fluctuation [2]. The cell containing the solution to be tested and a reference cell containing distilled water are placed in the same Mach–Zehnder interferometer sharing the same physical path length. The two photodetectors placed behind the last beam splitters measure the optical phase associated with the optical path length relative to the reference interferometric arm (l_1) for the signal interferometric arm l_2 for the solution and water, respectively. Since the optical paths for the solution and water have the same physical length, the error associated with the air temperature fluctuation is canceled if we compare the phase detected by the two photodetectors.

C.2   Precise Length Measurement

Optical interferometry can be used for length measurement at extremely high resolution. Figure C.2 illustrates the principle of the first generation of LIGO (laser interferometry for gravitational wave observatory) [3]. The gravitational wave detectors are designed to detect extremely small changes in space–time, on the order of 10^{-18} m or less in the length change, caused by the distortion in the gravitational field due to gravitational waves.

The detector is a Michelson interferometer where two interferometric arms are configured as a laser resonator. Initially, the relative arm length is adjusted so that the interference is totally destructive behind the beam splitter where the photodetector is placed. When a gravitational wave comes to the earth, it expands the space–time in one direction and contracts in the orthogonal direction. When this happens, the condition of the destructive interference breaks, and the photodetector senses light oscillating at the frequency of the

Fig. C.2 Highly sensitive
length measurement

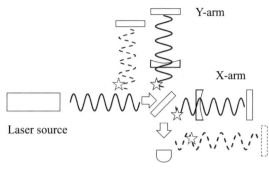

detected gravitational wave. Since the length change is extremely small, various tricks are made. One of them is to make the optical path as long as possible. By letting the laser beam travel back and forth inside the optical resonators, the optical path length is effectively increased by orders of magnitude.

Note that when the space–time is distorted, the length between the beam splitter and the end mirror and the wavelength of the laser light change for the same amount. Therefore, the number of waves between the beam splitter and the end mirror remains unchanged. However, since the time is distorted as well, the relative phase difference between the light beams in the two arms changes. In Fig. C.2, this effect is illustrated with the comparison of the solid and dashed waves where the space along the horizontal arm is expanded and the space along the other arm is contracted. Since the time is distorted in the same way (i.e., lengthen horizontally and shorten vertically) at the moment when a certain phase in the vertical path reaches the beam splitter the same phase, which causes the destructive interference before the disturbance, in the horizontal path does not reach the beam splitter. In this fashion, the condition of destructive interference breaks. The time distortion may be hard to understand. We can think of the effect as follows. According to special relativity, [22] the speed of light is constant. The speed of light is the length that light travels divided by the time taken for the travel. Since the ratio is constant, if the length increases by some percentage, so does the time.

References

1. J. Stone, J. Zimmerman (2001), Index of Refraction of Air [online], http://emtoolbox. nist.gov/Wavelength/Documentation.asp (Accessed August 21, 2022)
2. L. Woodside, S. Coppock, S. Yoshida, D. Norwood, Measurement of solution concentration by optical interferometry, American Phys. Soc., Annual APS March Meeting, March 3-7, Austin, TX, 2003
3. B. C. Barish, R. Weiss, LIGO and detection of Gravitational waves, Phys. Today 52 44–50, 1999.

Appendix D
Solving Helmholtz Equation

Substitution of Eq. (3.77) into Eq. (3.76) yields the following equation:

$$2i\,Q + 2k\frac{\partial P}{\partial z} + \left(Q^2 + k\frac{\partial Q}{\partial z}\right)r^2 = 0 \tag{D.1}$$

For Eq. (D.1) to hold for any r, it is necessary that the following two equations hold separately:

$$iQ + k\frac{\partial P}{\partial z} = 0 \tag{D.2}$$

$$Q^2 + k\frac{\partial Q}{\partial z} = 0 \tag{D.3}$$

To solve Eq. (D.2), put $Q(z)$ in the following form:

$$Q(z) = k\frac{\partial s/\partial z}{s} \tag{D.4}$$

By substituting Eq. (D.4) into Eq. (D.3), we find

$$\frac{\partial^2 s}{\partial z^2} = 0 \tag{D.5}$$

Equation (D.5) lets us put s with constants a and b in the following form:

$$s = az + b \tag{D.6}$$

Using Eq. (D.6), we can express $Q(z)$ in the following form:

$$Q(z) = k\frac{a}{az + b} \tag{D.7}$$

© The Editor(s) (if applicable) and The Author(s), under exclusive license to
Springer Nature Switzerland AG 2023
S. Yoshida. *Fundamentals of Optical Waves and Lasers*. Synthesis Lectures on Wave
Phenomena in the Physical Sciences, https://doi.org/10.1007/978-3-031-18188-7

Further, defining q as follows:

$$q(z) = \frac{k}{Q(z)} = k\frac{az+b}{ak} = z + \frac{a}{b} \tag{D.8}$$

we can write Eq. (D.7) in the following form:

$$q(z) = z + q_0 \tag{D.9}$$

Here $q_0 = a/b$ is a constant.

Use of Eq. (D.9) in Eq. (D.2) yields

$$P(z) = \int \frac{-iQ}{k}dz = \int \frac{-i}{q}dz = -i\int \frac{1}{z+q_0}dz = -i\ln(1 + \frac{z}{q_0}) \tag{D.10}$$

From Eq. (D.2), we find

$$Q(z) = -\frac{k\partial P(z)/\partial z}{i} = \frac{k}{z+q_0} \tag{D.11}$$

Substituting Eqs. (D.10) and (D.11) into Eq. (3.77), we obtain the following expression for $\Psi(r, z)$

$$\Psi = \Psi_0 e^{-i\left[-i\ln(1+\frac{z}{q_0}) + \frac{k}{2(z+q_0)}r^2\right]} = \Psi_0 e^{-\ln(1+\frac{z}{q_0})}e^{-i\frac{k}{2(z+q_0)}r^2} \tag{D.12}$$

Here the constant q_0 is potentially a complex number. For now, assume that q_0 is an imaginary number. It will become clear why we want to set q_0 in this fashion.

$$q_0 = i\frac{\pi w_0^2}{\lambda} \tag{D.13}$$

Here $\lambda = 2\pi/k$ is the wavelength and the meaning of w_0 will be discussed shortly.

Consider expressing $\ln(A + iB)$ (A and B are real numbers) in the following form:

$$\ln(A + iB) = \ln(re^{i\theta}) = \ln\left(\sqrt{A^2 + B^2}e^{i\theta}\right) = \ln\left(\sqrt{A^2 + B^2}\right) + \ln(e^{i\theta})$$
$$= \ln(\sqrt{A^2 + B^2}) + i\theta \tag{D.14}$$

Here $\theta = \tan^{-1}(B/A)$.

Using Eqs. (D.13) and (D.14), we can express the first exponential term on the right-hand side of Eq. (D.12) as follows:

$$e^{-\ln(1+\frac{z}{q_0})} = e^{-\ln\left(1 - i\frac{\lambda z}{\pi w_0^2}\right)} = e^{-\ln\left(\sqrt{1 + \left(\frac{\lambda z}{\pi w_0^2}\right)^2}\right) + i\tan^{-1}\left(\frac{\lambda z}{\pi w_0^2}\right)}$$

$$= e^{\ln\left(\sqrt{1 + \left(\frac{\lambda z}{\pi w_0^2}\right)^2}\right)^{-1}} e^{i\tan^{-1}\left(\frac{\lambda z}{\pi w_0^2}\right)} = \frac{1}{\sqrt{1 + \left(\frac{\lambda z}{\pi w_0^2}\right)^2}} e^{i\tan^{-1}\left(\frac{\lambda z}{\pi w_0^2}\right)} \tag{D.15}$$

Now consider the second exponential term on the right-hand side of Eq. (D.12). We put $q_0 = q_{0r} + iq_{0i}$ to see the effect of the real part of q_0 in the case it is a complex number.

$$e^{-i\frac{k}{2(z+q_0)}r^2} = e^{\frac{-ikr^2}{2}\frac{1}{(z+q_{0r})+iq_{0i}}} = e^{\frac{-ikr^2}{2}\left(-i\frac{q_{0i}}{(z+q_{0r})^2+q_{0i}^2} + \frac{z+q_{0r}}{(z+q_{0r})^2+q_{0i}^2}\right)}$$

$$= e^{\frac{-kr^2}{2}\left(\frac{q_{0i}}{(z+q_{0r})^2+q_{0i}^2}\right)} e^{\frac{-ikr^2}{2}\left(\frac{z+q_{0r}}{(z+q_{0r})^2+q_{0i}^2}\right)} \tag{D.16}$$

We can view the first exponential term of Eq. (D.16) as part of the amplitude of the paraxial wave. It indicates that the amplitude decreases quadratically with the radial distance from the z-axis. From the form of the function, we can see that the radial dependence of the amplitude is a Gaussian distribution. The denominator indicates that the width of the Gaussian distribution varies with z whereas its peak is always on the axis ($r = 0$). By setting $q_{0r} = 0$, we can make this term symmetric about $z = 0$.

The second exponential term on the last line of Eq. (D.16) represents the phase of the paraxial wave. Setting $q_{0r} = 0$ means that the phase at $z = 0$ is zero. In other words, the initial phase of the paraxial wave at $z = 0$ is null.

Now we find that setting the real part of q_0 to zero makes the z-dependence of the amplitude paraxial wave symmetric around $z = 0$ and setting the initial phase at $z = 0$ to zero, $\phi(0) = 0$. This setting is practically convenient, as we saw in Sect. 3.3.2. In the section below, we continue the discussion with q_0 as a pure imaginary number defined by Eq. (D.13).

First, denote the imaginary part of q_0 as z_0, i.e., $q_{0i} \equiv z_0$.

$$z_0 = \frac{\pi w_0^2}{\lambda} = \frac{kw_0^2}{2} \tag{D.17}$$

This quantity z_0 is known as the Rayleigh length (See Sect. 3.3.1). k on the right-hand side of Eq. (D.17) is the wavenumber, $k = 2\pi/\lambda$.

Putting $q_{0r} = 0$ and using Eq. (D.17), we can write Eq. (D.16) in a compact form:

$$e^{-i\frac{k}{2(z+q_0)}r^2} = e^{\frac{-kr^2}{2}\left(\frac{z_0}{z^2+z_0^2}\right)} e^{\frac{-ikr^2}{2}\left(\frac{z}{z^2+z_0^2}\right)} = e^{\frac{-z_0r^2}{w_0^2}\left(\frac{z_0}{z^2+z_0^2}\right)} e^{\frac{-ikr^2}{2}\frac{z^2}{z(z^2+z_0^2)}}$$

$$= e^{\frac{-r^2}{w_0^2\left(1+\left(\frac{z}{z_0}\right)^2\right)}} e^{\frac{-ikr^2}{2z\left(1+\left(\frac{z_0}{z}\right)^2\right)}} \tag{D.18}$$

By defining the following quantities:

$$R(z) = z\left(1 + \left(\frac{z_0}{z}\right)^2\right) \tag{D.19}$$

$$w(z) = w_0\sqrt{1 + \left(\frac{z}{z_0}\right)^2} \tag{D.20}$$

we can simplify the expression (D.18) as follows:

$$e^{-i\frac{k}{2(z+q_0)}r^2} = e^{\frac{-r^2}{w_0^2\left(1+\left(\frac{z}{z_0}\right)^2\right)}} e^{\frac{-ikr^2}{2z\left(1+\left(\frac{z_0}{z}\right)^2\right)}} = e^{-r^2\left(\frac{1}{w(z)^2} + \frac{ik}{2R(z)}\right)} \tag{D.21}$$

The right-hand side of Eq. (D.21) indicates that $w(z)$ is the value of radius r where the amplitude is $1/e$ of the peak value and that $R(z)$ is the radius of curvature of the wavefront. The quantity $w(z)$ is called the spot size of the Gaussian beam. The spot size and radius of curvature vary as a function of z according to Eq. (D.21).

Equation (D.20) indicates that the spot size takes the minimum value w_0 at $z = 0$ and increases as the wave travels in the positive or negative z-direction. The minimum spot size w_0 is called the beam waist size. Equation (D.19) tells us that the radius of curvature is infinity at $z = 0$ and $z \to \infty$, indicating that a paraxial wave's wavefront is flat at the beam waist position and infinitely far away from that position.

Combining Eqs. (D.15) and (D.21), we can express the $\Psi(r, z)$ of $S(r, z)$ term in Eq. (3.72) as follows:

$$\Psi(r, z) = \Psi_0 \frac{w_0}{w_0\sqrt{1 + \left(\frac{z}{z_0}\right)^2}} e^{i\tan^{-1}\left(\frac{z}{z_0}\right)} e^{-r^2\left(\frac{1}{w(z)^2} + \frac{ik}{2R(z)}\right)}$$

$$= \Psi_0 \frac{w_0}{w(z)} e^{-\frac{r^2}{w(z)^2}} e^{i\phi_G} e^{-i\frac{kr^2}{2R(z)}} \tag{D.22}$$

Here $\phi_G \equiv \tan^{-1}(z/z_0)$ is known as the Gouy phase. ϕ_G represents the on-axis phase reference to the initial phase at $z = 0$.

Index

© The Editor(s) (if applicable) and The Author(s), under exclusive license to   175
Springer Nature Switzerland AG 2023
S. Yoshida. *Fundamentals of Optical Waves and Lasers.* Synthesis Lectures on Wave
Phenomena in the Physical Sciences, https://doi.org/10.1007/978-3-031-18188-7

Printed in the United States
by Baker & Taylor Publisher Services